Institutional interplay

Institutional interplay: Biosafety and trade

Edited by Oran R. Young, W. Bradnee Chambers, Joy A. Kim and Claudia ten Have

TOKYO · NEW YORK · PARIS

© United Nations University, 2008

The views expressed in this publication are those of the authors and do not necessarily reflect the views of the United Nations University.

United Nations University Press
United Nations University, 53-70, Jingumae 5-chome,
Shibuya-ku, Tokyo 150-8925, Japan
Tel: +81-3-3499-2811   Fax: +81-3-3406-7345
E-mail: sales@hq.unu.edu   general enquiries: press@hq.unu.edu
http://www.unu.edu

United Nations University Office at the United Nations, New York
2 United Nations Plaza, Room DC2-2062, New York, NY 10017, USA
Tel: +1-212-963-6387   Fax: +1-212-371-9454
E-mail: unuona@ony.unu.edu

United Nations University Press is the publishing division of the United Nations University.

Cover design by Rebecca S. Neimark, Twenty-Six Letters

Printed in Hong Kong

ISBN 978-92-808-1148-3

Library of Congress Cataloging-in-Publication Data

Institutional interplay : biosafety and trade / edited by Oran R. Young ... [et al.].
    p. cm.
  Includes bibliographical references and index.
  ISBN 978-9280811483 (pbk.)
  1. International trade—Environmental aspects. 2. Commercial policy—
Environmental aspects. 3. Biotechnology—Government policy.
I. Young, Oran R.
HF1379.I5464   2008
363.19—dc22   2007052015

We dedicate this volume to Konrad von Moltke

Contents

Figures and tables ix

Foreword x
    A. H. Zakri

Contributors xii

Part I: Introduction to the issues 1

1 Institutional interplay and the governance of biosafety 3
    W. Bradnee Chambers, Joy A. Kim and Claudia ten Have

2 Global biosafety governance: Emergence and evolution 19
    Aarti Gupta

Part II: Institutional interplay and its application to biosafety and trade 47

3 Analysing biosafety and trade through the lens of institutional interplay 49
    Heike Schroeder

4 Overlapping regimes: The SPS Agreement and the Cartagena Biosafety Protocol 71
    Are K. Sydnes

5 Disentangling the interaction between the Cartagena Protocol and the World Trade Organization 94
   Sebastian Oberthür and Thomas Gehring

Part III: Conclusion 129

6 Deriving insights from the case of the WTO and the Cartagena Protocol... 131
   Oran R. Young

Part IV: Remembering Konrad von Moltke 159

7 The WTO as an environmental agency 161
   Steve Charnovitz

8 Additional tributes to Konrad von Moltke 192

Index 199

Figures and tables

Figures
3.1 IDGEC's research agenda 52
5.1 A causal mechanism of institutional interaction 99
6.1 Dimensions of institutional interplay 139

Tables
2.1 Importer decisions: The WTO SPS Agreement and the Cartagena Protocol 31
3.1 Institutional interplay: Categories of structural linkage 58
3.2 Principles of regimes governing transboundary movement in GMOs 60
3.3 Objectives of regimes governing transboundary movement in GMOs 65

Foreword

Bridging theory and practice have long been a foundation of the United Nations University's mission. New understanding and new knowledge are the key to enlightening decision makers about the information they require to progress and work together towards solving the most pressing issues of our time. This book is an excellent example of such a bridge.

For many years Oran Young and his colleagues have led international networks of scholars to think about how institutions work, why they are important and ultimately how they can be improved to better serve the goals they were put in place to achieve. His research and writings on institutional interplay are an excellent example of this cutting-edge work. In 1996 when the UNU-IAS was established, its initial work on Sustainable Development Governance also led it to similar topics. It quickly picked up the theme of "interlinkages", a concept that reflected the international diplomatic world's effort to manage the aftermath of over two decades of intense treaty-making and institution-building in the field of environment and sustainable development. This period left diplomats and practitioners facing a deeply fragmented and overlapping governance system.

Since 1996 both these concepts of interplay and interlinkages have stimulated a growing body of literature and have even spurred several schools of thought on the topic. However, this work lacks a better understanding of how these concepts and theories can be applied practically in the real world. This book is an attempt to fill this gap, and biosafety is a fitting example.

Biosafety is a topic that implicates many different organizations and institutions that underpin the international trade, development and environment regimes. On the trade side, biosafety is a major concern of the World Trade Organization and its agreements on Sanitary and Phytosanitary measures and Technical Barriers to Trade. On the development side, the Codex Alimentarius, the International Plant Protection Convention and the International Office of Epizootics have become recognized as providing the world's major food safety standards. On the environment and sustainable development side, the 1999 Cartagena Protocol on Biosafety has become the basis for ensuring that the trade in "living modified organisms" does not threaten biodiversity conservation and sustainable use. Balancing the sometimes competing interests of these regimes has not always been easy and has generated a great deal of uncertainty and intense debate about how these regimes can cooperate.

Reconciling these regimes will be a major challenge for the international community in the years to come. The importance of this timely book is that it provides decision makers with the knowledge that will allow them to better understand the nature and behaviour of the regimes and ultimately how they can work together to create a more coherent international system.

Konrad von Moltke himself was an innovator in the study of regime overlap and was among the first to pioneer ideas of how to bring greater coherence to international governance. It is in recognition of his many contributions to this field of study that we dedicate this book to his memory.

A. H. Zakri
Director, UNU-IAS

Contributors

W. Bradnee Chambers is Senior Programme Officer at the United Nations University Institute of Advanced Studies (UNU-IAS), Yokohama, Japan. He was a convening lead author of the Millennium Ecosystem Assessment and for the United Nations Environment Programme's GEO-4 *Chapter 8: Interlinkages: Governance for Sustainability*. He is a Senior Legal Fellow at the Centre for International Sustainable Development Law in Montreal, Canada. He specializes in public international law and international relations, and works on environmental treaty and international economic issues. His current research concentrates on finding ways to improve cooperation among multilateral environmental agreements based on an interlinkages approach.

Steve Charnovitz is Associate Professor of Law at the George Washington University Law School, Washington DC, USA. Prior to joining the faculty, Professor Charnovitz practised law for six years at Wilmer Hale in Washington, DC. From 1995 to 1999 he was the director of the Global Environment and Trade Study (GETS) at Yale University, USA, and from 1991 to 1995 he was the policy director of the US Competitiveness Policy Council. Professor Charnovitz serves on the editorial boards of the *American Journal of International Law*, the *Journal of International Economic Law*, the *Journal of Environment & Development*, and the *World Trade Review*. A collection of his essays, *Trade Law and Global Governance*, was published in 2002.

Thomas Gehring is Professor of International Relations at the Faculty of Social Sciences, Economics and Business Administration of Otto-Friedrich

University in Bamberg, Germany. He received his PhD from the Free University in Berlin and was a Jean-Monnet Fellow of the European University Institute in Florence, Italy. Professor Gehring has written widely on international institutions and European integration. He is the author of *Dynamic International Regimes. Institutions for International Environmental Governance* (1994). And, more recently, he is also the co-editor with Sebastian Oberthür of *Institutional Interaction in Global Environmental Governance: Synergy and Conflict among International and EU Policies* (2006).

Aarti Gupta is Assistant Professor with the Environmental Policy Group at Wageningen University, The Netherlands. She focuses on global environmental and risk governance, the role of science in policy-making, and the interaction of environmental regimes with trade rules. Her recent work has been on global–national linkages in the governance of agricultural biotechnology. Prior to coming to Wageningen, Professor Gupta was a post-doctoral researcher at Columbia University's Center for Science, Policy and Outcomes, and a Global Environmental Assessment Fellow at Harvard University's Belfer Center for Science and International Affairs. She received her PhD from Yale University's Department of Forestry and Environmental Studies in 2001.

Joy A. Kim is Senior Policy Analyst on Trade and Environment at the Trade and Agriculture Directorate of the Organisation for Economic Co-operation and Development, Paris, France. Her research focuses on integrated policy-making in the area of environmental governance. Dr Kim works and publishes on issues related to institutional interplay, interlinkages between multilateral environmental agreements, climate change and sustainable development, as well as trade liberalization and environmental goods and services. She is a contributing author to UNEP's GEO-4 *Chapter 8: Interlinkages: Governance for Sustainability*.

Sebastian Oberthür is Academic Director of the Institute for European Studies (IES) in Brussels, Belgium. With a PhD in political science from the Free University of Berlin and a background in international law, Dr Oberthür focuses on issues of international and European environmental governance, with an emphasis on institutional issues. Recently, he co-edited *Institutional Interaction in Global Environmental Governance: Synergy and Conflict among International and EU Policies* with Thomas Gehring (2006). Previously he worked as a Senior Research Fellow with Ecologic, the Institute for International and European Environmental Policy in Berlin.

Heike Schroeder is Tyndall Research Fellow in the Environmental Change Institute at the Oxford University Centre for the Environment, UK. She holds a PhD from the Free University of Berlin, Germany. Until early 2007, Dr Schroeder was the Executive Officer of the Institutional Dimensions of Global Environmental Change (IDGEC) research project. At this time she was also a post-doctoral researcher at the Bren School of Environmental Science and Management, University of California, Santa Barbara, USA.

Are K. Sydnes is Associate Professor in Societal Security and Environment at the University College of Tromsø, Norway. He received his PhD from the University of Tromsø, where he also has held positions as post-doctoral research fellow and Associate Professor in political science.
Dr Sydnes has published widely on issues of natural resource management, most recently as a co-author with Tore Henriksen and Geir Hønneland of *Law and Politics in Ocean Governance: The UN Fish Stocks Agreement and Regional Fisheries Management Regimes* (2006).

Claudia ten Have is Associate Fellow at the United Nations University Institute of Advanced Studies (UNU-IAS), Yokohama, Japan. She works on institutional aspects of governance for sustainable development, including interplay and interlinkages. She is a lead author for UNEP's GEO-4 *Chapter 8: Interlinkages: Governance for Sustainability*. She holds a PhD in Public Administration. Before coming to Japan, Dr ten Have was a Research Fellow at the South African Institute of International Affairs (SAIIA) and lectured in the International Relations Department at the University of the Witwatersrand, Johannesburg.

Oran R. Young is Professor and Co-Director of the Program on Governance for Sustainable Development at the Bren School of Environmental Science and Management at the University of California, Santa Barbara, USA. Since 2006 Professor Young has been the Chair of the Science Committee of the International Human Dimensions Programme (IHDP) on Global Environmental Change. Prior to this he was the Chair of the Scientific Steering Committee of the research project on the Institutional Dimensions of Global Environmental Change (IDGEC). Professor Young's scientific work encompasses both basic research focusing on collective choice and social institutions and applied research specific to international environmental governance and the Arctic as an international region. He is the author or co-author of over 20 books and numerous scholarly articles.

Part I
Introduction to the issues

1
Institutional interplay and the governance of biosafety

W. Bradnee Chambers, Joy A. Kim and Claudia ten Have

1. Introduction

International institutions and the consequences of their interplay are emerging as a major agenda item for research and policy. As governments enter into an ever increasing number of international agreements, so questions arise about the overlap of issues, jurisdiction and membership. Of particular interest to practitioners and analysts is how this mélange of institutions at the international level intersects and interrelates to influence and affect the content, operation, performance and effectiveness of a specific institution, as well as the functioning of the overall global governance context.

Interplay – here understood to refer to the phenomenon where one institution intentionally or unintentionally affects another (King 1997) – is set to increase, as additional international institutions are created and as existing institutions co-evolve through international and national implementation. The question of how parties to international negotiations and concluded agreements should deal with this situation has given rise to three interrelated analytical themes:

1. What are the links and pathways of inter-institutional influence and interaction? In short, what is the *process of interplay*?
2. What implications does this interaction hold for the interests of the stakeholders, for the formation, operation and implementation of the specific institution, for its performance and effectiveness, and for

Institutional interplay: Biosafety and trade, Young, Chambers, Kim and ten Have (eds), United Nations University Press, 2008, ISBN 978-92-808-1148-3

the overall global governance context? In other words, what are the *politics of interplay*?
3. A line of enquiry linked with political and policy efforts to strengthen overall governance focuses on identifying and stimulating *interlinkages among institutions* (legal, normative, operational and functional) in order to reduce institutional conflict and resource-draining duplication. Put differently, in what ways can issues and processes across institutions be strategically linked and arranged to reinforce each other? Or how can we *manage interplay*?

Although discourse on regimes and institutions over the past 30 years has made significant contributions to our understanding of the role and functioning of regimes and associated institutions on particular issue areas at the international level (see, for example, Young 1982; Krasner 1983; Haas et al. 1993; Underdal 1995; Levy et al. 1995; Hasenclever et al. 1997), the study of interplay among institutions and across issue areas at that level is still relatively new and under-explored. Oran R. Young and his colleagues pioneered efforts to lay out the research agenda for conceptual work on such institutional interplay in the mid-1990s (Young 1996; King 1997; Young et al. 1999) as part of the Institutional Dimensions of Global Environmental Change (IDGEC) Programme, a long-term international research project under the auspices of the International Human Dimensions Programme on Global Environmental Change (IHDP). This initial conceptual work was significantly extended by Young in the years to follow (Young 2002; Young 2006), as well as by scholars linked to the Fridtjof Nansen Institute in Norway – particularly Olav Schram Stokke (2001), Kristin Rosendal (2001a, 2003) and Regine Andersen (2002) – and by the German scholars Sebastian Oberthür and Thomas Gehring (Oberthür 2001; Oberthür and Gehring 2003, 2006).

This volume brings together these different scholars to apply their various insights on interplay to the issue of biosafety governance. As is detailed below and in the next chapter, biosafety – that is, measures to minimize negative impacts of biotechnology – is an issue that is relevant to many institutions and thus offers an excellent case study for exploring and applying interplay in practical terms. The purpose of this volume is not so much to add to the already extensive literature on biosafety governance per se,[1] but to use the issue of biosafety and the institutions involved in it – chiefly the World Trade Organization (WTO) and the Cartagena Protocol on Biosafety to the Convention on Biological Diversity (CBD) – as a window through which to assess what we understand conceptually and practically about institutional interplay. To date no study brings together different scholars and their various contributions to the study of interplay in this manner.[2] In the remainder of this chapter we provide a short introduction to the study of interplay and the issue of biotechnology and trade, followed by an outline of the book.

2. The study of institutional interplay

With the rising density of institutions at the international level has come greater attention to the issue of their interaction and interplay. Given that institutions have a definite spatial remit (in terms of issue, jurisdiction and membership),[3] so-called "boundary problems" are central to institutional interplay (Moss 2004: 2). The boundaries at stake here relate to political responsibilities and social spheres of influence. The crux in the study of institutional interplay is that for the most part institutions are not self-contained entities and so the effectiveness of specific institutions depends often not only on their own features but also on their interactions with other institutions (Young et al. 1999: 60). In addition, although such interplay is a common and familiar feature in the domestic context, where procedures have evolved over time to manage linkages, how this interaction and its effects can be managed at the international level in the absence of a central governing authority presents an important concern for research and practice (Young 2002: 9; see research on interlinkages reported by the United Nations University since 1999).

Oran R. Young's early work mapped the analytical landscape for the study of institutional interplay, which was differentiated in terms of four types of linkage (Young et al. 1999: 62–64). The first type is *functional linkages*, in the sense that the operation of one institution directly influences the effectiveness of another through some substantive connection among the activities involved. Secondly, *political linkages* are involved when actors actively seek to link and/or integrate two or more institutions. Young et al. further showed that interplay occurs along both a horizontal and a vertical axis. *Vertical linkages* cut across levels of social organization, whereas *horizontal linkages* are found among institutional arrangements operating at the same level of social organization. Young also differentiated between interplay that occurs when institutional arrangements are *embedded* in and informed by overarching principles and practices; when arrangements are *nested* by design within functionally and/or geographically broader regimes; when arrangements are the result of deliberate *clustering* of several regimes across functional and/or geographical borders; and when arrangements simply *overlap* largely unintentionally (Young 1999: 165–172).

Much of the analytical energy of the Norwegian scholars went one step deeper, into understanding the causes and effects of this interplay; there were also early steps to develop a theoretical foundation for interplay.

Olav Schram Stokke of the Fridtjof Nansen Institute sought to understand how and why interplay occurs. He identified "causal pathways of interplay" and distinguished among four modalities: *diffusion*, where one institution may influence the material content of another; *political spillover*, where the interests or capabilities of one institution influence the operation of another; *normative interplay*, where the rules upheld in one institution conflict with or reinforce those established in another; and *operational interplay*, where the activities of separate institutions are deliberately coordinated to avoid normative conflict or wasteful duplication. In this way Stokke differentiated among normative, political and operational interplay (Stokke 2000). Casting these differentiations within the study of the literature of the effectiveness of regimes to begin deeper theoretical work on interplay, Stokke later made a distinction between *utilitarian interplay* (which is incentive driven), *normative interplay* (which is commitment driven) and *ideational* interplay (which is learning driven) (Stokke 2001: 12).

Another scholar attached to the Fridtjof Nansen Institute, Kristin Rosendal, looked at effectiveness by investigating the conditions in which interplay has disruptive, as opposed to supportive, effects. Rosendal proposed differentiating on the one hand between general *norms* and specific *rules*, and on the other between whether these principles and rules are diverging or compatible (Rosendal 2001a: 97). In this way four situations of interplay can occur: rules and norms are compatible (a synergistic situation); norms diverge but rules are compatible (a relatively synergistic situation); norms are compatible but rules diverge (a potentially problematic situation); and norms and rules diverge (a problematic situation). She added to this by proposing a conceptual differentiation between *core* and *secondary* aspects of regimes, and pointed out that situations where core aspects differ offer greater scope for conflict than do those with differences between secondary aspects. Similarly, she differentiated between *regulatory* and *programmatic* rules, and argued that the likelihood of conflict is greater where regulatory rules diverge than where programmatic rules diverge (Rosendal 2001a: 98–101).

The Norwegian scholars applied their concepts to various case studies of interplaying institutions such as regional and global regimes managing fish stocks (Stokke 2000); to the overlap between the Intergovernmental Forum on Forests, the United Nations Framework Convention on Climate Change (UNFCCC) and the CBD (Rosendal 2001b); to the interaction between the CBD and the WTO Agreement on Trade-Related Aspects of Intellectual Property Rights (TRIPS) on the issue of access to genetic resources (Rosendal 2003); and to the interplay among the CBD, TRIPS and the International Treaty on Plant Genetic Resources for Food and Agriculture (ITPGRFA) in the management of plant genetic diversity in agriculture (Andersen 2002). Through examining the interplay of the CBD, TRIPS and ITPGRFA. Regine Andersen, another scholar at the Fridtjof Nansen Institute, underlined the importance of factoring the *time dimension* into understanding international institutional interplay. She pointed out that institutions' stages of development had different implications for how they in turn interact with and affect other institutions (Andersen 2002). Different types of interplay are at work during the negotiation of an agreement, compared with its early and with its later, more advanced stages of implementation.

The German scholars Sebastian Oberthür and Thomas Gehring too directed their research efforts to deepening understanding of the causal mechanisms that drive institutional interaction and the circumstances in which institutional interplay produces synergistic as opposed to disruptive or conflict-ridden outcomes. Rather than issue-based case studies of institutional interaction (e.g. on the governance of plant genetic resources between the CBD, TRIPS and the ITPGRFA), Oberthür and Gehring sought a generalized framework of analysis of the phenomenon of institutional interaction (Oberthür and Gehring 2003). In this way their approach and methodology differ significantly from those of the Norwegian scholars. Oberthür and Gehring's analysis focuses on disaggregating interplay into its simplest form as *unidirectional influence flows between source and target institution.* Their point of departure was to identify flows among a given set of institutions (in their case the WTO, the UNFCCC, the CBD, the Convention on International Trade in Endangered Species, the UN Fish Stocks Agreement, and institutions for the protection of the North-East Atlantic) and so to untangle incidents of institutional interaction that are intentional from interaction that is non-intentional, unilaterally induced or requiring consent from the target institution, as well as interaction that is conflictual or synergistic. In this way they arrived at two broad types of interaction: what might be called "soft interplay", or in their terminology "cognitive interaction" (through a change in perception, which could occur when one institution serves as a model for another or when one institution requests another to change), and "hard interplay", or in their terms "interaction with a stick" (where the source institution forces the target institution to change, which can occur through jurisdictional delimitation, and through filtering down new preferences through broader or nested institutional arrangements).

Despite this considerable collection of typologies and classifications of interplay, it can be argued that the study field is fragmented and that deeper analysis is needed of *how institutional interplay affects global governance.* Limited progress has been made on rooting the study of interplay theoretically in this regard (see Stokke 2001). Also, the distance between the concepts developed to study interplay and their empirical and policy application has been, and continues to be, rather substantial. The United Nations University has been part of efforts since the late 1990s to bridge this gap through research focused on the "interlinkages" among environment and sustainable development governance institutions (see United Nations University 1999) and on the linkages between climate governance and other multilateral regimes (Chambers 1998, 2001).

To address the gap between theory and practice, and to open the discussion on interplay to a deeper consideration of its theoretical implications, this volume looks at the problématique of institutional interplay through a focused case study, namely the global governance of biosafety. This is a new step in this larger and ongoing research process on interplay.[4] The volume brings together some of the aforementioned scholars to consider the case of biosafety governance from their specific conceptual framework – Heike Schroeder, attached to IDGEC, Oran R. Young's transnational collaborative research project on the institutional dimensions of global environmental change; Are K. Sydnes of the University of Tromsø in Norway applying the perspectives of his Norwegian colleagues; and Sebastian Oberthür and Thomas Gehring examining biosafety governance through their disaggregated approach to institutional interplay. The objective of this research project is to profile the contributions these different scholars have made to our understanding of institutional interplay, to identify what theoretical ground remains to be covered, and to collect the insights offered by the different scholars' approaches for our understanding of what influences the effectiveness of governance per se, and the global governance framework on biosafety in particular.

3. The case of biosafety governance

Rapid advances have taken place over the past 30 years in the field of biotechnology.[5] These sophisticated techniques and their commercialization are set to have an immense impact on agricultural production and food sciences, pharmaceuticals and diagnostic processes in medicine, as well as the development of new industrial products. Although biotechnology encompasses a range of techniques and sub-fields and there is little controversy about many aspects of its traditional applications, the science to manipulate the genetic structures of cells – resulting in genetically modified organisms (GMOs) and therefore the ability to develop transgenic micro-organisms, plants and animals and derivative commercial products – has captivated public attention and become the target of intense debates.

On the one hand, transgenic food crops and animals are welcomed for their promise of higher yields, improved nutrition or resistance to pests and diseases; as are genetically altered micro-organisms associated with breakthroughs in new medical therapies and in new fuels, materials (for example, bioplastics) and industrial processes (for example, waste treatment) that have the potential to be cleaner and more resource efficient. On the other hand, given the newness of modern biotechnology and the limited knowledge of how its products may behave and evolve in interaction with the natural world in the long term, genetic engineering and particularly the associated global trade of its products have raised a range of environmental, health, social and ethical concerns and strong calls for adequate safety measures. Policy makers and regulators at both the national and international levels have therefore been faced with the complex challenge of developing appropriate legislation and risk assessment systems to secure the safety of globally traded biotechnology products, and at the same time balancing this with ensuring unhindered market access as stipulated by binding WTO rules and obligations.

From the start biosafety – understood here to encompass measures, policies and procedures to minimize and eliminate potential environmental and human health risks resulting from biotechnology and its products (in particular GMOs) – was set to become a knotty global governance matter. Global efforts at biosafety rule-making are complicated by a sharp divide in the values and expectations of major stakeholders regarding transgenic products, with GMO-exporting countries (both developed and developing) backed by a powerful and growing biotech industry colliding with (potential) GMO-importing countries sensitive and responsive to strong public and consumer opinion against genetically engineered products.

Moreover, given the range of concerns that biosafety governance is required to address, it is little surprise that rule-making has emerged in numerous institutions. Early steps towards a regulatory response to GMOs were taken in the 1980s in the United Nations Food and Agriculture Organization (FAO), the World Health Organization (WHO), the United Nations Industrial Development Organization (UNIDO), the Organisation for Economic Co-operation and Development (OECD), and the United Nations Environment Programme (UNEP). Of relevance too are a number of international instruments dealing with adjacent aspects of plant and animal health, as well as food safety, that had already been in place for many years, including the International Office of Epizootics' animal health standards, the International Plant Protection Convention's plant health standards, and the food safety standards of the FAO/WHO's Codex Alimentarius Commission.

The two key institutions, however, that came to the fore in the 1990s as the pivots in the emerging global governance architecture on biosafety were the Cartagena Protocol on Biosafety, concluded under the Convention on Biological Diversity (CBD) between 1996 and 2000, and the World Trade Organization (WTO) and its associated trade agreements and standard-setting instruments.

Although WTO agreements and instruments do not explicitly deal with GMOs or the issue of biosafety, a number of agreements contain provisions relevant to the transboundary movement of traded goods and are thus of direct relevance to traded GMO products. The WTO agreement of immediate relevance is the 1994 Sanitary and Phytosanitary (SPS) Agreement, which prevents national sanitary (human and animal health) and phytosanitary (plant health) measures from becoming non-tariff barriers to trade. Of partial relevance is the WTO's Technical Barriers to Trade (TBT) Agreement (also 1994), which regulates technical standards in cases not covered by the SPS.

The Cartagena Protocol on Biosafety, in contrast, deals exclusively with GMOs. Negotiations towards the Protocol were launched as a result of provisions stipulated during the formation of the CBD in the early 1990s. The objectives of the CBD are "the conservation of biological diversity, the sustainable use of its components and the fair and equitable sharing of the benefits arising out of the utilization of genetic resources" (CBD Article 1). While the Convention was being drafted negotiators recognized that biotechnology could make an important contribution towards achieving the CBD's objectives, if developed and used within adequate safety measures for the environment and human health. This led to the decision to consider procedures to secure the safe transfer, handling and use of GMOs resulting from biotechnology that may have adverse effects on the conservation and sustainable use of biological diversity (see Article 19.3 of the CBD). This in turn led to the negotiation of the Cartagena Protocol to the Convention on the issue of biosafety. In accordance with the precautionary approach contained in Principle 15 of the Rio Declaration on Environment and Development, the objective of the Protocol is to establish practical rules and procedures for the safe transfer, handling and use of GMOs – or, in the Protocol's language, "living modified organisms" (LMOs) – that result from modern biotechnology, with a specific focus on the transboundary movement of such items. Since it specifically addresses the transboundary movements of GMO products, the legally binding Protocol has direct implications for the international trade in GMOs and related products. It establishes differentiated procedures for GMOs to be intentionally introduced into the environment (e.g. seeds, micro-organisms, fish) and for transgenic commodities intended for direct use as food, as animal feed or for processing (e.g. corn, grain, soya, canola, tomatoes). These procedures are intended to provide importing countries with key data to make informed decisions on whether or not to accept the GMO imports, and on mechanisms and institutions to handle GMOs in a safe manner. Incidentally, the Protocol does not cover all GMO products. Pharmaceuticals are not covered, nor are products derived from GMOs such as cooking oil from GM corn or ink from GM soya.

That the WTO and the Protocol would emerge as the axis of global biosafety governance was clear early on. They overlap in their means to achieve their objectives – that is, both seek to create international standards that are implemented through binding trade measures. They also overlap largely in their membership – there are 141 parties to the Protocol and 150 to the WTO (as of June 2007).[6] Yet they differ significantly in their objectives. The WTO is about market access and views GMO trade from the exporter perspective, aiming to ensure that products are treated in a non-discriminatory manner, save some particular exceptions. The Protocol, in contrast, is anchored in precaution and, through the advance informed agreement (AIA) procedure, enables importers to put in place and operate risk assessment and management procedures that seek to minimize GMO risks.

Four issues proved contentious during the negotiation of the Protocol: the scope of the Protocol (in particular, whether it would cover GMOs for direct use as food, as feed or for processing); the decision matrix and role of the precautionary principle; the Protocol's relationship to other agreements; and the question of liability and redress (Cosbey and Burgiel 2000). These issues proved contentious primarily because of their direct interplay with the WTO system. Although agreement was ultimately reached on each of these questions – in some cases with a significant measure of creative and diplomatic ambiguity[7] – and countries on both sides of the debate praised the Protocol for accommodating WTO rules, debate continues on whether the Protocol's provisions complement or compete with those of the WTO (see, for example, Phillips and Kerr 2000; Rivera-Torres 2003). Questions also continue about which of the two would prevail should disputes be brought forward for adjudication.[8] In addition, analytical attention to both regimes is set to continue as both evolve further through future global rule-making, redefinition and national implementation. As Aarti Gupta points out in Chapter 2, although the emerging global biosafety framework has been carefully assembled, it remains unclear how the components of this rapidly expanding set of global rules actually interact with and influence each other.

Biosafety governance clearly is relevant to numerous institutions. Given this volume's aim of testing different conceptual approaches to interplay, the global governance of biosafety is of particular interest and relevance from three angles. First, insights can be drawn from a historical analysis of the negotiation and formation of the Protocol to see the effect of the WTO's rules and obligations on the Protocol's negotiating parties and their decisions. Secondly, insights into current and ongoing developments in the institutional interplay in the regulation of the transboundary movement of GMOs can give important cues for the future biosafety regime, as well as for other instances of institutional interaction at the trade/environment intersection. And, lastly, a study of the interplay between the WTO and the Protocol can yield important insights into how better coordination among the links between the two regimes could strengthen overall biosafety governance effectiveness while reconciling the legitimate interests of trade, biosafety and other sectors.

4. Overview of this book

This volume proceeds in four parts. Chapter 2 by Aarti Gupta completes Part I by setting the stage for utilizing the emerging global governance framework for biosafety to assess different conceptual approaches to institutional interplay. Gupta sketches the institutional and political context within which calls to regulate the safe transboundary movement of GMOs emerged, and then details the rules and obligations under the SPS Agreement of the WTO and other related agreements and standard-setting mechanisms. She also provides a full account of the provisions and workings of the Cartagena Protocol on Biosafety. Gupta's historical overview of the Protocol negotiation process and the different negotiating groups shows not only the complex interest structure of parties to the negotiations but also their clear and constant awareness of the Protocol's functional relationship with the WTO. The underlying message is that no new effort at rule-making for traded GMOs could be made without reference to WTO obligations and provisions, given that the vast majority of countries participating in the Cartagena negotiations were also party (or future party) to the binding agreements of the WTO. The subsequent decisions by Cartagena negotiators to patch over potentially conflictual and contentious issues – such as the operationalization of precaution, as well as the issue of the Protocol's relationship to other agreements – through creative ambiguity can therefore be seen as resulting directly from interplay with the WTO. Given that both the Protocol and the WTO are set to evolve further, and by implication to continue to interact, Gupta also identifies three linkage areas that interplay scholars ought to watch closely for cues on how the global governance of biosafety might develop: the negotiation of information-sharing obligations for the agricultural commodity trade; the transmission of global biosafety rules to the domestic context through capacity-building and dispute settlement decisions; and the impact of the evolving membership and alliance groups of the Protocol.

In Part II, theory meets practice as the interplay between the biosafety and trade regimes is reviewed by scholars of interplay. In Chapter 3, Heike Schroeder, using insights into institutional interplay generated by the IDGEC project, differentiates between issue-based, goal-based and power-based political interplay, as well as between horizontal and vertical interplay, and between functional and political interplay. She then differentiates between forms of dependence among institutions (reciprocal versus unidirectional) and describes how institutions can be structurally linked into embedded, nested, clustered and overlapping arrangements. She finds that the interplay between the biosafety and the trade regimes is horizontal and functional and is likely to continue to be a reciprocal relationship.

In Chapter 4, Are K. Sydnes applies the insights generated at the Fridtjof Nansen Institute and looks at interplay in biosafety governance through the prism of "overlapping" institutions as adapted and defined by Kristin Rosendal. He discusses how overlap is dealt with by institutions and identifies a range of different means, including codification, international law, political interpretation and negotiation, deliberate coordination, "turf wars" and "forum shopping". He considers in what circumstances interplay turns malignant or benign and adopts the distinctions made by Rosendal in this regard, namely differentiation between core aspects and secondary aspects of the regime, and between regulatory and programmatic rules. Combining Rosendal's categories with Stokke's normative, political and operational interplay, Sydnes makes three propositions: first, that the core aspects and regulatory rules of regimes are more politically sensitive than other types of overlap; secondly, that normative interplay is most benign in cases where the core aspects and regulatory rules of regimes are compatible; and, thirdly, that programmatic regulations are more benign to operational interplay between regimes than other substantive or operational aspects of institutional overlap.

In Chapter 5, Sebastian Oberthür and Thomas Gehring provide a detailed and updated version of their methodology for studying institutional interaction. In contrast to the previous two chapters, which both take a holistic approach to the study of interplay, Oberthür and Gehring identify specific cases of interaction in a single source institution and a single target institution and a unidirectional causal mechanism connecting the two. Expanding on their earlier work (2003), they identify four causal mechanisms in total. The first two affect the decision-making of an institution, namely *cognitive interaction* and *interaction through commitment*.

The second two affect an institution's implementation and effectiveness, namely *behavioural interaction* and *impact-level interaction*. They find that an institution can influence others in four ways: through diffusing new information, knowledge or ideas; through its commitments affecting the preferences of actors in other institutions; by inducing behaviour changes within the issue area governed by another institution; and through the direct side effects of its impacts on the ultimate target of protection. The Cartagena Protocol on Biosafety claimed regulatory authority over biosafety in the mid-1990s, and the parties' proclaimed commitment at the start of the negotiations to address biosafety under the CBD umbrella prevented the WTO from reclaiming regulatory rights over it later. Oberthür and Gehring also find that, although the Protocol displayed surprising ability at the beginning to secure its rights to assume regulatory space, it was negotiated under and continues to function in the "shadow of the WTO". This can be seen particularly in the Cartagena provisions on risk assessment and socio-economic considerations, as well as in the ambiguity in its relationship with other agreements, notably those of the WTO.

In Part III, Oran R. Young provides reflections and conclusions on the chapters in Part II and their insights into the study of interplay and its application to the case of biosafety and trade. Young points out the limitations of the proliferation of interplay taxonomies, which have little to offer a deeper theoretical understanding of interplay. Instead of the current catalogue-like list of interplay types, Young proposes concentrating on two differentiations only, namely whether interplay is *intended or not*; and whether interplay is *shallow or deep*. Shallow interplay here refers to superficial interaction, whereas deep interaction goes far beyond the operational interaction to encompass principles, norms and values. Young argues that interplay that is deep, intentional and conflictual is the most contentious and difficult to address. In this way Young significantly extends Rosendal's work. Such interplay is likely to be the focus of future interplay studies. Young shows that the interplay between the WTO and its related institutions and the Protocol can be interpreted to be deep, intentional and conflictual, raising significant questions about the future evolution of the biosafety regime.

The final part departs in form from the rest of the volume, and a note on this is in order at this stage. This part of the book is a special tribute to Konrad von Moltke. During the planning of this volume we had invited Konrad von Moltke to contribute a chapter reviewing the Protocol from the trade perspective. We could think of no better scholar for this question given von Moltke's pioneering and inspiring work at the trade/environment interface over the past two decades. Despite many other commitments, he cheerfully agreed. Some months later, in May

2005, however, we were immensely saddened to hear of his untimely passing.

As editors we are grateful to Steve Charnovitz, a long-time friend, for agreeing at very short notice to contribute an exploratory chapter on "The WTO as an Environmental Agency" instead. Given the WTO's vast membership, economic clout and binding rules and obligations, it is typically thought of as the dominant or (using Oberthür and Gehring's vocabulary) "source" institution affecting the content, operations and effectiveness of other institutions. Certainly, the previous chapters show how the negotiations for the newer biosafety regime were influenced and circumscribed by the regulatory space already occupied by the WTO. However, the WTO is not immune to or cut off from interactive effects from other institutions. In fact, as Steve Charnovitz shows in his fresh and provocative chapter, the WTO too is the target of influence from various other institutions, and this has affected the content, the operation and, some would argue, the effectiveness of the WTO. In short, the WTO has in fact "endogenized" some of the influences of interplay emanating from the environmental side.

Notes

During the preparation of this chapter Claudia ten Have was a Japan Society for the Promotion of Science (JSPS) Fellow at UNU-IAS.

1. See, for example, Phillips and Kerr (2000), Gupta (2000), Bail et al. (2002), Safrin (2002), Brack et al. (2003) and Rivera-Torres (2003).
2. An exception is the recent edited volume by Oberthür and Gehring (2006), which brings together a number of case studies of interplay at the international and European Union level.
3. "Institutions" are here understood in their broadest sense as "sets of rules, decision-making procedures and programmes that define social practices, assign roles to the participants in these practices, and guide interactions among the occupants of individual roles" (Young 2002: 5).
4. In December 2006, IDGEC held its Synthesis Conference in Bali, Indonesia, where the work thus far on "interplay" was reviewed. See Sebastian Oberthür and Thomas Gehring's conference paper, "Interplay: Exploring Institutional Interaction"; available at ⟨http://www2.bren.ucsb.edu/~idgec/responses/Sebastian%20Oberthuer%20et%20al%20-%20Interplay.doc⟩ (accessed 2 July 2007).
5. The term "biotechnology" refers to any technological application that uses biological systems or living organisms, or derivatives thereof, to make or modify products or processes for specific use. Traditional biotechnology includes fermentation techniques as well as plant- and animal-breeding techniques such as hybridization. In modern biotechnology, researchers can take a single gene from a plant or animal cell and insert it in another plant or animal cell to produce a desired characteristic, such as a plant resistant to a particular pest. In the Cartagena Protocol on Biosafety (see Article 3), modern biotechnology means the application of:

a. in vitro nuclei techniques, including recombinant DNA and direct injection of nucleic acid into cells or organelles, or
   b. fusion of cells beyond the taxonomic family
   that overcome natural physiological reproductive or recombination barriers and that are not techniques used in breeding and selection.

6. As is shown in greater detail in Chapter 2 by Aarti Gupta, important countries that have *not* ratified the Cartagena Protocol on Biosafety include the United States, Canada, Australia and Singapore. The United States is also not party to the Protocol's parent agreement, the Convention on Biological Diversity.
7. See Chapter 2 for detail on the choice of the term "living modified organism" (LMO) instead of "genetically modified organism" (GMO), as well as on the wording of the preamble.
8. The most prominent case in this regard is the May 2003 complaint by the United States, Argentina and Canada to the WTO about the de facto moratorium on the approval of new GMOs, as well as a number of marketing and import bans (so-called "safeguard measures"), in certain European Union countries. See Baumüller et al. (2006).

REFERENCES

Andersen, Regine (2002), "The Time Dimension in International Regime Interplay", *Global Environmental Politics* 2(3): 98–117.

Bail, Christopher, Robert Falkner and Helen Marquard, eds (2002), *The Cartagena Protocol on Biosafety: Reconciling Trade in Biotechnology with Environment and Development?* London: RIIA/Earthscan.

Baumüller, Heike, Knirie Søgaard and Yvonne Apea (2006), "Overview of the WTO Biotech Dispute and the Interim Ruling", International Centre for Trade and Sustainable Development, Geneva, March.

Brack, Duncan, Robert Falkner and Judith Goll (2003), *The Next Trade War? GM Products, the Cartagena Protocol and the WTO*. London: RIIA.

Cartagena Protocol (2000), *Cartagena Protocol on Biosafety to the Convention on Biological Diversity: Text and Annexes*. Montreal: Secretariat of the Convention on Biological Diversity; available at ⟨http://www.cbd.int/doc/legal/cartagena-protocol-en.pdf⟩ (accessed 5 July 2007).

Chambers, W. Bradnee, ed. (1998), *Global Climate Governance: Inter-linkages between the Kyoto Protocol and other Multilateral Regimes*. Tokyo: United Nations University Press.

Chambers, W. Bradnee, ed. (2001), *Inter-Linkages: The Kyoto Protocol and the International Trade and Investment Regimes*. Tokyo: United Nations University Press.

Convention on Biological Diversity (1992), text available at ⟨http://www.biodiv.org/convention/convention.shtml⟩ (accessed 2 July 2007).

Cosbey, Aaron and Stas Burgiel (2000), *The Cartagena Protocol on Biosafety: An Analysis of Results. An IISD Briefing Note*, Winnipeg, Canada: International Institute for Sustainable Development; available at ⟨http://www.iisd.org/pdf/biosafety.pdf⟩ (accessed 3 July 2007).

Gupta, Aarti (2000), "Governing Trade in Genetically Modified Organisms: The Cartagena Protocol on Biosafety", *Environment* 42(4): 23–33.

Haas, Peter M., Robert O. Keohane and Marc A. Levy, eds (1993), *Institutions for the Earth: Sources of Effective International Environmental Protection*. Cambridge, MA: MIT Press.

Hasenclever, Andreas, Peter Mayer and Volker Rittberger (1997), *Theories of International Regimes*. New York: Cambridge University Press.

King, Leslie A. (1997), "Institutional Interplay: Research Questions. A Report for Institutional Dimensions of Global Change, International Human Dimensions Programme on Global Environmental Change", Draft, University of Vermont, September; available at ⟨http://www2.bren.ucsb.edu/~idgec/publications/idgecscience/InstitutInterplay.pdf⟩ (accessed 3 July 2007).

Krasner, Stephen D., ed. (1983), *International Regimes*. Ithaca, NY: Cornell University Press.

Levy, Marc A., Oran R. Young and Michael Zürn (1995), "The Study of International Regimes", *European Journal of International Relations* 1: 26–330.

Moss, Timothy (2004), "Regional Sustainable Development as a Cross-Sectoral Task", discussion paper presented at the REGIONET workshop "Cross Fertilization and Integration of Results of REGIONET", Brussels, 14–16 January 2004; available at ⟨http://www.iccr-international.org/regionet/docs/ws4-moss.pdf⟩ (accessed 3 July 2007).

Oberthür, Sebastian (2001), "Linkages between the Montreal and Kyoto Protocols – Enhancing Synergies between Protecting the Ozone Layer and the Global Climate", *International Environmental Agreements: Politics, Law and Economics* 1(3): 357–377.

Oberthür, Sebastian and Thomas Gehring (2003), "Investigating Institutional Interaction: Towards a Systematic Analysis", paper presented at the International Studies Association Convention, Portland, Oregon, 12 February to 1 March.

Oberthür, Sebastian and Thomas Gehring, eds (2006), *Institutional Interaction in Global Environmental Governance: Synergy and Conflict among International and EU Policies*. Cambridge, MA: MIT Press.

Phillips, Peter B. P. and William A. Kerr (2000), "The WTO versus the Biosafety Protocol for Trade in Genetically Modified Organisms", *Journal of World Trade* 34(4): 63–75.

Rivera-Torres, Olivette (2003), "The Biosafety Protocol and the WTO", *Boston College International & Comparative Law Review* 26: 263–323.

Rosendal, G. Kristin (2001a), "Impacts of Overlapping International Regimes: The Case of Biodiversity", *Global Governance* 7(1): 95–117.

Rosendal, G. Kristin (2001b), "Overlapping International Regimes. The Case of the Intergovernmental Forum on Forests (IFF) between Climate Change and Biodiversity", *International Environmental Agreements: Politics, Law and Economics* 1: 447–468.

Rosendal, G. Kristin (2003), "Interacting International Institutions: The Convention on Biological Diversity and TRIPS: Regulating Access to Genetic Resources", paper presented at the International Studies Association Convention, Portland, Oregon, 12 February to 1 March.

Safrin, Sabrina (2002), "Treaties in Collision? The Biosafety Protocol and the World Trade Organisation Agreements", *American Journal of International Law* 96(3): 606–628.

Stokke, Olav Schram (2000), "Managing Straddling Stocks: The Interplay of Global and Regional Regimes", *Ocean & Coastal Management* 43: 205–234.

Stokke, Olav Schram (2001), "The Interplay of International Regimes: Putting Effectiveness Theory to Work", FNI Report 14, Fridtjof Nansen Institute, Oslo.

Underdal, Arild (1995), "The Study of International Regimes", *Journal of Peace Research* 32: 113–119.

United Nations University (1999), *Inter-Linkages: Synergies and Coordination between Multilateral Environmental Agreements*, Report of the conference available at ⟨http://www.geic.or.jp/interlinkages/docs/UNUReport.PDF⟩ (accessed 2 July 2007).

Young, Oran R. (1982), "Regime Dynamics: The Rise and Fall of International Regimes", *International Organization* 36(2): 277–297.

Young, Oran R. (1996), "Institutional Linkages in International Society: Polar Perspectives", *Global Governance* 2: 1–24.

Young, Oran R. (1999), *Governance in World Affairs*. Ithaca, NY: Cornell University Press.

Young, Oran R. (2002), *The Institutional Dimensions of Environmental Change: Fit, Interplay, and Scale*. Cambridge, MA: MIT Press.

Young, Oran R. (2006), "Vertical Interplay among Scale-dependent Environmental and Resource Regimes", *Ecology and Society* 11(1): 27; available online at ⟨http://www.ecologyandsociety.org/vol11/iss1/art27/⟩ (accessed 3 July 2007).

Young, Oran R., with Arun Agrawal, Leslie A. King, Peter H. Sand, Arild Underdal and Merrilyn Wasson (1999), *Science Plan: Institutional Dimensions of Global Environmental Change*, IHDP Report No. 9. Bonn: International Human Dimensions Programme on Global Environmental Change.

2

Global biosafety governance: Emergence and evolution

Aarti Gupta

1. Introduction

Although it is the focus of increasing attention, global governance of biosafety is relatively recent and still in a stage of evolution. Concern with safe use of biotechnology can be traced back to the 1970s, when gene splicing first occurred in the United States (US). Until the mid-1990s, however, legally binding biosafety regulation was concentrated at the national level and regionally through European Union (EU) directives. It is only since the mid-1990s that a flurry of multilateral negotiations have been establishing the foundations for an emerging global biosafety governance framework. Global regimes establishing rules and norms for trade in genetically modified organisms (GMOs) as well as their safe transfer and use are now coming into force and becoming part of international law.

The global rule-making effort most directly focused on GMOs is the Cartagena Protocol on Biosafety, negotiated under the Convention on Biological Diversity (CBD) and in force since 2003. The other key global regime relevant for biosafety is the Agreement on the Application of Sanitary and Phytosanitary Measures (SPS) of the World Trade Organization (WTO), which was concluded and came into force in 1994. The Cartagena Protocol and the SPS Agreement contain rules and processes to regulate trade in GMOs, hence they are central to an emerging global biosafety governance architecture. A number of other agreements and institutions are also potentially relevant, including the WTO's Agreement

Institutional interplay: Biosafety and trade, Young, Chambers, Kim and ten Have (eds), United Nations University Press, 2008, ISBN 978-92-808-1148-3

on Technical Barriers to Trade (TBT) and the Codex Alimentarius Commission of the Food and Agriculture Organization and the World Health Organization (FAO/WHO), a global food safety standard-setting body that is now debating international safety standards for food produced through use of genetic engineering.

This emerging governance framework has to contend with a wide range of concerns (including ecological, human health, social and ethical) associated with the use of modern biotechnology in sectors such as agriculture and medicine. The governance challenge is made more complex by the fact that the existence, nature and manageability of risks associated with modern biotechnology remain deeply contested. The emerging global biosafety framework has been the focus of much analytical attention (see, for example, Isaac and Kerr 2003; Safrin 2002; Coleman and Gabler 2002; Buckingham and Phillips 2001). Despite this, it remains unclear how the components of this rapidly expanding set of global rules interact with and influence one another. This is partly because the various regimes are still evolving, and their obligations are still being interpreted and/or expanded within a variety of global forums as well as through national implementation.

A key influence on this process of global biosafety regime evolution is the ongoing transatlantic conflict over trade in GMOs. The past decade has witnessed an expanding use of biotechnology in agriculture, with resultant "transgenic" crops spreading worldwide. This has been accompanied by an escalating trade conflict between the United States and the European Union over GMOs, fuelled by a de facto EU moratorium (in place between 1999 and late 2003) on domestic approvals of transgenic crops (Brack et al. 2003). The European Union has the most stringent set of regional regulations governing the use of GMOs in agriculture, which have been developed in a political context of growing consumer opposition to transgenic crops in the past decade. The global biosafety governance framework is being strongly influenced by this transatlantic GMO conflict. This is, then, a crucial juncture for global biosafety governance, with the transatlantic conflict shaping both its evolution and potential interactions between its component parts.

In such circumstances, it is timely to systematically analyse, as this volume seeks to do, whether the rules and norms of global biosafety governance are evolving in complementary or contradictory ways, and with what implications. This chapter sets the stage for this question to be considered from a variety of different perspectives by analysing the current "state of play" in global biosafety governance. It describes the negotiating history and objectives of the key global regimes, and points to areas of potential regime interplay.

In the next section I examine the emergence of a global biosafety framework, by outlining the negotiating rationales and objectives underlying the WTO's SPS Agreement, other related global regimes and the Cartagena Protocol on Biosafety. I then consider the evolution of the global biosafety governance framework, by reviewing recent developments within these global regimes that are likely to be important for regime interplay. In the conclusion, I highlight trends to consider in analysing regime interplay and discuss the importance of biosafety linkages for multilateral environmental governance in general.

2. The emergence of a global biosafety governance framework

Developing global rules to regulate trade in substances that might pose ecological or human health risks received sustained attention in the 1990s as a result of the transatlantic "beef hormone" conflict. The conflict derives from a 1989 ban by the European Union on imports of beef treated with certain growth hormones. The official rationale for the ban was the risk posed to human health from the hormones.

The United States and Canada, key exporters of beef to the European Union, alleged that the ban was not based on scientific evidence of harm and viewed it as a non-tariff barrier to trade motivated by protectionism. The negotiation of the SPS Agreement can be traced to this conflict, as a further step in the world trade regime's attempt to harmonize national health and safety regulations to prevent them from becoming disguised barriers to trade.[1]

Concurrent with this, the safety of traded products has also been the subject of separate, free-standing multilateral environmental agreements (MEAs). These include international treaties regulating trade in hazardous waste as well as banned and restricted pesticides (the Basel and Rotterdam conventions, respectively[2]). However, separate MEAs regulating trade in risky substances exist mainly where the hazardous nature of the traded substance is not disputed. Both the Basel and Rotterdam conventions call for the "prior informed consent" of an importing country before trade in hazardous wastes or banned and restricted chemicals can occur. Prior to negotiation of the Cartagena Protocol on Biosafety, it was unusual to have a separate global treaty regulating trade in a substance whose hazardous nature remained contested. For such cases, the disciplines of the SPS Agreement provide the de facto governance regime, as with trade in hormone-treated beef. Yet, pushed by developing countries that feared the "dumping" of novel and potentially hazardous

GMOs in their territories, the Cartagena Protocol on Biosafety was negotiated to allow for the "advance informed agreement" of an importing country prior to trade in certain GMOs. It is thus particularly instructive to examine how this global regime's obligations and rules relate to those of the SPS Agreement. Each global regime's negotiating rationale and objectives are discussed further below.

2.1. The SPS Agreement and related global regimes: Negotiating rationale and objectives

The explicit objective of the SPS Agreement is to prevent national sanitary (human and animal health) and phytosanitary (plant health) measures from becoming non-tariff barriers to trade. The Agreement seeks to prevent ad hoc protectionism through establishing guidelines and processes upon which to base national health and safety regulations, while allowing for legitimate context-specific differences in appropriate levels of safety. This is reflected in the SPS Agreement's opening paragraph, which states that "no Member should be prevented from adopting or enforcing measures necessary to protect human, animal or plant life or health, subject to the requirement that these measures are not applied in a manner which would constitute an arbitrary or unjustifiable discrimination between members where the same conditions prevail or a disguised restriction on international trade" (SPS Agreement 1994: Preamble).

Two kinds of adverse impacts on trade from domestic health and safety regulations are feared: impacts on the competitiveness of traded goods, and impacts on market access. Competitiveness concerns arise if lower environmental, health and safety standards in particular locales confer a competitive advantage on such locales over those with higher standards (Esty and Geradin 1997). Such an imbalance can lead to a flight of industry to areas with lower health and safety standards or to the much-debated "race to the bottom" in environmental health and safety standards across jurisdictions (Revesz 1992; Klevorick 1996). Market access concerns arise when (higher) national health and safety standards have the effect of impeding access to such markets. Such standards could be either higher than those in other jurisdictions or merely easier for domestic producers to meet.

In the case of GMOs, and especially in the context of the ongoing transatlantic trade conflict, the primary issue is restricted market access, where importing country regulatory standards are more stringent than elsewhere, with the effect of restricting or denying access to its market. This concern underpins the WTO case brought by the United States in 2003 against the European Union's regulatory approach to GMOs. The United States, supported by Canada and Argentina, viewed the European Union's GMO regulations (and its de facto moratorium on new approvals, lifted in 2004) as inconsistent with its WTO obligations (WTO 2003; see also Isaac and Kerr 2003). As this case made clear, the central challenge for the WTO and particularly for its SPS Agreement is how to permit "necessary" national sanitary and phytosanitary measures while preventing "arbitrary" restrictions on trade (and to agree on ways to systematically assess and agree upon concepts such as necessity or arbitrariness).

The SPS Agreement seeks to meet this challenge by establishing rules and procedures upon which to base national health and safety regulations – with harmonization of domestic regulatory approaches seen as desirable to prevent negative impacts on trade. One route to harmonization is adoption of international standards; hence the SPS Agreement encourages member states to "further the use of harmonized sanitary and phytosanitary measures between Members, on the basis of international standards, guidelines and recommendations developed by the relevant international organizations" (SPS Agreement 1994: Preamble). Countries are encouraged to voluntarily adopt the Codex Alimentarius Commission's food safety standards, the International Office of Epizootics' animal health standards and the International Plant Protection Convention's plant health standards. In the case of GMOs, if countries adopt safety standards and procedures for genetically modified food (currently being developed by the Codex Commission), this would be seen as consistent with their SPS obligations.

Where relevant international standards do not exist or where countries choose to maintain higher standards, the SPS Agreement seeks to harmonize the bases for maintaining higher standards (SPS Agreement 1994; see also Marceau and Trachtman 2002). Thus, the agreement allows for higher standards to be maintained if "the relevant international standards ... are not sufficient to achieve [a state's] appropriate level of sanitary or phytosanitary protection". What constitutes an "appropriate" level can also be determined by a member state as long as any measure is "applied only to the extent necessary to protect human, animal or plant life or health, is based on scientific principles and is not maintained without sufficient scientific evidence" (SPS Agreement 1994: Article 2).

Thus, at the heart of the SPS Agreement is a requirement that higher national standards must have clear scientific justification. The SPS Agreement does make an exception in cases where there is scientific uncertainty about adverse effects on human, plant or animal health and safety. Thus, Article 5.7 of the SPS Agreement states that:

> In cases where relevant scientific evidence is insufficient, a Member may provisionally adopt sanitary and phytosanitary measures on the basis of available

pertinent information ... In such circumstances, Members shall seek to obtain the additional information necessary for a more objective assessment of risk and review the sanitary or phytosanitary measure accordingly within a reasonable period of time. (SPS Agreement 1994: Article 5.7)

Such precautionary measures are, however, to be maintained only on a provisional basis while additional objective scientific data on risk are sought. Although the motivation is to keep "arbitrary" non-scientific considerations out of the decision calculus, the SPS Agreement does allow for "relevant economic factors" to be considered in the risk assessment. These include "the potential damage in terms of loss of production or sales in the event of the entry, establishment or spread of a pest or disease; the costs of control or eradication in the territory of the importing Member; and the relative cost-effectiveness of alternative approaches to limiting risk" (SPS Agreement 1994: Article 5.3). Thus, socio-economic factors that can be included in a risk assessment under the SPS Agreement are linked to potential economic damage from sanitary or phytosanitary harm. However, this excludes socio-economic considerations not directly linked to sanitary or phytosanitary harm, such as public acceptability or consumer opposition, considerations that do nonetheless influence domestic debate and regulatory choices about GMOs. The emphasis on science-based decision-making in the SPS Agreement has been much analysed in recent years. The potential for objective science to reduce conflicts is at the centre of scrutiny within the SPS Agreement itself, with a series of disputes over domestic SPS measures (and their scientific justifications) coming before the WTO dispute settlement mechanism in recent years (for an overview, see Christoforou 2000; for a critique of the view that "objective" science can resolve political conflicts, see Gupta 2004; Sarewitz 2000; Jasanoff 1998a, 1998b).

It remains an open question so far whether the WTO's TBT Agreement might be applicable in cases where domestic GMO regulations are based not upon safety concerns but rather upon other "legitimate" objectives, as permitted under the TBT. The TBT Agreement covers domestic technical standards and regulations in cases where the SPS Agreement does not apply (TBT Agreement 1994; Marceau and Trachtman 2002). It prohibits discrimination between "like products", an issue that has long dogged GMO debates, given disagreement over whether GMOs are "like" their non-GMO counterpart products. One recent argument explored by analysts is whether consumer opposition to GMOs alone might make them *unlike* their non-GMO counterparts (regardless of safety), hence permitting domestic regulation of GMOs under the TBT Agreement, as long as a legitimate purpose can be shown. On this latter point, however, it has not yet been put to the test whether consumer right-to-know, for example, might constitute a legitimate purpose for technical regulations and standards under the TBT Agreement, with its non-exhaustive list of what such legitimate purposes might be. In general, non-safety-related arguments have not yet been made by countries in justifying their domestic GMO regulations in global forums (Appleton 2000; Baumüller 2003).

In an important development, a debate about non-science or "other legitimate factors" that can be considered in food safety assessments (including genetically modified food) is under way within the Codex Alimentarius Commission (Codex 2003; see also Boutrif 2003; Skogstad 2001; Cosby n.d.). A Codex "Statement of principle regarding the role of science in the Codex decision-making process and the extent to which other factors are taken into account" attempts to strike a balance between "science-based decisions" and other legitimate factors. Although what these are is not spelled out, a reference is made to fair trade practices, as well as other legitimate factors that might be "accepted worldwide" rather than be applicable only to a particular jurisdiction (Codex n.d.). Although conflicts over interpreting such concepts will continue, the discussion has important implications for the trade regime.

In the particular case of GMOs, a Codex Ad Hoc Intergovernmental Task Force on Foods Derived from Biotechnology has sought to elaborate on its understanding of legitimate factors underpinning risk regulation, in developing a set of principles for risk analysis of genetically modified (GM) foods. These principles include reference to risk management measures, including controversial issues such as tracing, labelling and post-approval monitoring of GMOs, as potentially appropriate tools for domestic risk regulation. This mirrors arguments made by the European Union in other global forums (including at the WTO in the beef hormone conflict) and by developing countries (in the Cartagena Protocol on Biosafety) that appropriate handling and use, including capacity for appropriate handling, are important to consider in domestic risk regulation. The Codex principles on risk analysis do, however, make reference to the need for compatibility with the WTO, in particular with the SPS and TBT agreements (Codex 2003; also Covelli and Hohots 2003).

Although this is an important development, whether the adoption of Codex standards and procedures will aid in reducing conflict over regulatory approaches to GMOs is unclear. Again, the beef hormone experience is instructive here. In that case, Codex safety standards did exist for five of the hormones in question, yet these standards were adopted not by consensus (the normal method of functioning in Codex) but by a narrow margin of victory in a secret ballot requested by the United States: 33 governments approved the standards, 29 opposed them and 7 abstained (Kastner and Pawsey 2002). The European Union's subsequent argument in the WTO's beef hormone dispute settlement process, that the Codex standards were not based on consensus and hence the safety assessment contained therein was contested, was dismissed as irrelevant by the WTO panel, given that the SPS Agreement does not require that international standards be adopted by consensus. This may have important implications for global biosafety governance, since ongoing controversial debates within Codex over threshold levels and safety assessment procedures for GM foods may make consensus difficult to attain here as well. This throws into question whether global safety standards and guidelines, such as those of the Codex Alimentarius Commission, offer a way to reduce conflict and harmonize domestic regulations (as envisioned by the SPS Agreement) or whether they are inextricably caught up within the very same conflicts, with seemingly technical safety standards mirroring ongoing political conflicts in parallel global, regional and national forums. It also remains unclear whether Codex guidelines have the same status under the SPS Agreement that Codex standards (which are fixed targets) have.

Not surprisingly, similar disputes over appropriate procedures for national regulatory decisions have dogged the negotiation and evolution of the Cartagena Protocol on Biosafety as well. Heated debates have centred on, first, the Protocol's obligations to share information and solicit importing country agreement prior to trade in certain GMOs; and, second, the criteria upon which importing country agreement can be based. The Protocol's negotiating rationale and key obligations are discussed next.

2.2. The Cartagena Protocol on Biosafety: Negotiating rationale and objectives

After protracted negotiations over four years (1996–2000) and a temporary collapse of the negotiations in 1999 in Cartagena, Colombia (where the Protocol was to have been finalized), the Cartagena Protocol on Biosafety was concluded in Montreal in January 2000. It came into force on 11 September 2003 after 50 countries had ratified it. The first Meeting of the Parties to the Protocol was held in Kuala Lumpur, Malaysia, in February 2004, when key elements were agreed on to facilitate its implementation.

The Protocol was negotiated under the 1992 Convention on Biological Diversity (CBD) whose three-fold objective is conservation, sustainable use and the sharing of benefits from biological diversity. As a means of achieving these goals, the Convention calls on parties to "consider the need for and modalities of a protocol setting out appropriate procedures, including, in particular, advance informed agreement, in the field of the safe transfer, handling and use of any living modified organism resulting from biotechnology that may have adverse effect on the conservation and sustainable use of biological diversity" (1992: Article 19.3). This provision, which laid the groundwork for a biosafety protocol to be negotiated under the CBD, was the outcome of extensive debate over whether global regulation of GMOs was necessary (Rajan 1997; McConnell 1996; Gupta 2000a, 2000b). GMOs are called "living modified organisms" (LMOs) in the Protocol, in a language change already evident in the CBD provision above. This is a striking example of the conflict over whether GMOs should be subject to global regulation. The United States, at the forefront of research on transgenic crops at the time, pushed for the change to "living" modified organism to deflect attention away from "genetic" modification as the focus of regulatory attention (Rajan 1997). The US position was that genetic engineering did not pose unique hazards and did not need to be singled out for separate global regulation. This argument is in line with its domestic approach, which regulates GMOs under existing laws. Similarly, the United States pushed for use of the term "advance informed agreement" instead of the more commonly used "prior informed consent" because the latter is associated in the global realm with trade in hazardous substances such as waste and banned chemicals.

Five negotiating alliances shaped the development of global biosafety rules during negotiation of the Cartagena Protocol, especially in the last phase from 1999 to 2000. These were: the Miami Group of agricultural exporting countries (including Argentina, Australia, Canada, Chile, the United States and Uruguay); the European Union; the so-called Like-Minded Group of developing countries; Eastern and Central European countries in transition; and the so-called Compromise Group, consisting of members of the Organisation for Economic Co-operation and Development that were not part of the Miami Group or the European Union (such as Japan, Mexico, Norway, Singapore, South Korea, Switzerland and New Zealand).

By 1999, Miami Group countries were at the forefront of producing and commercializing transgenic crops; the United States was in the lead, followed by Argentina, Canada and Australia. This group thus represented the concerns of potential GMO-exporting countries in negotiating the Protocol, with an interest in minimizing exporter obligations to share information and solicit importer consent prior to GMO trade. As primary producers of transgenic crops, the group's key concern was to minimize restrictions on the bulk agricultural commodity trade – of crops such as soybean, maize and cotton – of which a growing percentage is now transgenic.[3]

In contrast, the European Union negotiated the Protocol from a potential GMO-importer perspective. The European Union was also operating in a political environment of increasing domestic opposition to transgenic crops, which has grown more intense in the past half-decade. The third negotiating alliance, the Like-Minded Group of developing countries, was the most ardent initial supporter of a biosafety protocol. These countries were concerned about the entry of potentially novel hazards into their territories and their lack of capacity to manage such hazards (Egziabher 1999; Bail et al. 2002). Central and East European countries voiced similar concerns with regard to their limited capacity to monitor the entry and safe use of LMOs, but they also supported EU positions in many cases, given the prospect of EU integration and need for harmonization with EU regulatory approaches in the future. The fifth negotiating alliance, the Compromise Group, reflected a mix of the concerns voiced by the Miami Group and the European Union, including as it did important agricultural importing countries such as Japan, as well as European leaders in biotechnology research such as Switzerland.

In addition to these negotiating alliances, a Global Industry Coalition of agricultural, food and pharmaceutical companies supported the Miami Group in pushing for a "workable" protocol with a narrow scope (Global Industry Coalition 1999). In contrast, an informal coalition of environmental and consumer safety advocates strongly supported developing-country demands for the broadest possible scope of the protocol's biosafety rules to facilitate oversight over all categories of traded GMOs (Greenpeace 1999; Third World Network 2000). A wide range of complex issues faced countries as they negotiated the Biosafety Protocol and many detailed analyses of the negotiating process and the perspectives of different groups now exist (see, in particular, Bail et al. 2002; Gupta 2000a, 2000b; Falkner 2000).

The centrepiece of the finalized Cartagena Protocol is an obligation on exporting countries to solicit the advance informed agreement of an importing country prior to the transfer of LMOs intended for deliberate release into the environment. Other categories of LMOs, such as those transferred for contained use (in research laboratories, for example), those intended for direct consumption as food or feed or for processing (agricultural commodities), or processed products deriving from LMOs (such as soybean oil produced from transgenic soya), do not require prior information-sharing and agreement from an importing country. For these categories of LMOs, exporting countries have an obligation to share certain information simultaneously with a transfer, although the nature and extent of these information-sharing obligations vary. LMO-based pharmaceuticals are excluded from the Cartagena Protocol altogether, as long as they are being addressed in other appropriate international forums (Cartagena Protocol 2000: Articles 4–7).

On the critical issue of exporter obligations for agricultural commodities (LMOs intended for food, feed or processing – called LMO-FFPs), developing countries lost the battle to have advance informed agreement apply to this category, which constitutes the bulk of GMOs entering international trade. Instead, an alternative procedure requires exporting countries to notify the Biosafety Clearing-House (the centralized institutional mechanism to share information about LMOs) of domestic approvals of LMOs within 15 days of the approval being granted, i.e. before the LMO has even been planted domestically in the country of production. This provides potential importing countries with the opportunity to consider, in advance of an LMO variety entering international trade, whether they want to assess its risks to their environment or human health and/or to restrict its import. If so, countries can inform the Biosafety Clearing-House of their decision within a set time frame. Although decisions must be based upon a risk assessment, countries may take import-restrictive decisions in the face of scientific uncertainty about harm posed by an LMO, as discussed further below.

There are important information-sharing obligations on exporting countries during shipments of agricultural commodities that contain transgenic varieties. This issue proved particularly controversial in the final hours of negotiating the Protocol in 2000. The conflict centred on what kind of information was required to accompany bulk commodity shipments containing LMO varieties, once these were under way. The European Union demanded that documentation accompanying agricultural commodity shipments with transgenic varieties clearly state the identity and unique characteristics of each LMO in a shipment. This demand was rejected by the Miami Group because it would have mandated segregation and tracking of LMO varieties through the agricultural commodity chain. The compromise reached in the final minutes of negotiations on the Protocol calls for documentation accompanying commodity shipments to state only that they "may contain" LMOs (rather than specifying which ones or giving additional information). Parties to the Protocol were required, however, to elaborate on these information requirements within two years of the Protocol coming into force (that is, by September 2005). This issue is now at centre stage of the Protocol's evolution, as discussed in section 3 below.

On the second key issue of the legitimate bases for domestic (possibly import-restrictive) decisions, the Protocol mandates that importer decisions about LMO transfers are to be based upon a scientific risk assessment. As stated in the Protocol, risk assessments are to be "carried out in a scientifically sound manner ... taking into account recognized risk assessment techniques" (Cartagena Protocol 2000: Article 15).

Furthermore, the Protocol allows (as does the SPS Agreement) precautionary restrictions on trade in cases of scientific uncertainty about the extent of harm posed by an LMO. Finally, the Protocol also allows certain socio-economic factors to be considered in importer decisions, as demanded by developing countries. Table 2.1 summarizes key provisions on precaution and socio-economic factors in both the SPS Agreement and the Cartagena Protocol on Biosafety, in order to facilitate comparison.

The interpretation of these clauses (and their relationships to each other) has been the focus of much analytical attention (Safrin 2002; Brack et al. 2003; Buckingham and Phillips 2001; Gupta 2001; Phillips and Kerr 2000). When the Cartagena Protocol on Biosafety was concluded, it was hailed as the first regime to operationalize the precautionary principle in a multilateral environmental agreement, and hence a milestone in global environmental governance. It is certainly a step forward for those who support precautionary decisions in biosafety and multilateral environmental agreements (MEAs). However, rather than operationalizing "the" precautionary principle, the Protocol's language on precaution was the outcome of last-minute negotiations and represents a mix of existing formulations. The main provisions on precaution, Articles 10.6 and 11.8 of the Protocol, take as their starting point the prior existence of a risk assessment. In this, they are consistent with the SPS Agreement. However, in contrast to the SPS Agreement, the Protocol's language on precaution does not include a time frame within which precautionary restrictions must be reviewed. It thus allows more flexibility to countries to restrict imports in the face of scientific uncertainty about the harm posed by an LMO.[4]

Regarding socio-economic factors, the Protocol's provision (see Table 2.1) links the permissible socio-economic impacts of traded GMOs to potential impacts on biodiversity. It also requires that socio-economic considerations be consistent with other international obligations (such as the WTO). Although it has not yet been subjected to legal interpretation, this clause excludes taking into account general social concerns associated with LMO trade, such as the impact on traditional livelihoods or dependence on patented seed, concerns voiced by developing countries during the Protocol's negotiation. Such concerns, even if legitimate to consider from a developing-country perspective, would almost certainly run foul of the WTO. However, a key question that remains open is whether it is permissible to take into account socio-economic considerations such as a lack of capacity to segregate traded LMOs or the ability to monitor safe handling and use. The ability to appropriately handle LMOs could be considered a legitimate risk management issue that countries need to take into account in deciding whether to permit LMO imports.

Table 2.1 Importer decisions: The WTO SPS Agreement and the Cartagena Protocol

Basis for importer decisions	WTO SPS Agreement	Cartagena Protocol on Biosafety
Precaution	"In cases where relevant scientific evidence is insufficient, a Member may provisionally adopt sanitary or phytosanitary measures on the basis of available pertinent information … In such circumstances, Members shall seek to obtain the additional information necessary for a more objective assessment of risk, and review the sanitary or phytosanitary measure accordingly within a reasonable period of time." (SPS Agreement 1994: Article 5.7)	"Lack of scientific certainty due to insufficient relevant scientific information and knowledge regarding the extent of the potential adverse effects of a living modified organism on the conservation and sustainable use of biological diversity in the Party of import, taking also into account risks to human health, shall not prevent the Party from taking a decision, as appropriate, with regard to the import of the living modified organism … in order to avoid or minimize such potential adverse effects." (Cartagena Protocol: Articles 10.6, 11.8)
Socio-economic factors	"Relevant economic factors" that can be considered within a risk assessment are "the potential damage in terms of loss of production or sales in the event of the entry, establishment or spread of a pest or disease; the costs of control or eradication in the territory of the importing Member; and the relative cost-effectiveness of alternative approaches to limiting risks". (SPS Agreement 1994: Article 5.3)	The Protocol does not include socio-economic factors within the parameters of a risk assessment. Instead, countries may (separate from the risk assessment) "take into account, consistent with their international obligations, socio-economic considerations arising from the impact of living modified organisms on the conservation and sustainable use of biological diversity". (Cartagena Protocol: Article 26.1)

As noted earlier, a similar discussion is under way within the Codex Alimentarius Commission.

Given the multiple potential interpretations of the Protocol's decision criteria (including vis-à-vis the WTO), the Protocol's relationship to other international agreements such as the trade regime was explicitly on the agenda during its negotiation. To ensure that WTO obligations would not be superseded by the Protocol, the Miami Group of countries argued that the Protocol should contain a "savings clause" or a provision stating that nothing in the Protocol affects a country's obligations under other international agreements. Other groups, and especially the European Union, opposed inclusion of such a clause. These groups argued that a savings clause would negate the purpose of negotiating a biosafety protocol and would subordinate it to the WTO. This was one of the last issues to be resolved prior to finalization of the Protocol. The compromise reached does not include a savings clause in the operative articles of the Protocol, as desired by the Miami Group. However, it does include explicit language in the preamble about the relationship between the Protocol and other international agreements. Although this has sparked debate over whether the preamble has sufficient legal standing vis-à-vis operative articles, it is generally agreed that, given a conflict over operational provisions, the preamble is examined to ascertain the intentions of countries negotiating a treaty.

However, the three paragraphs in the preamble that address the relationship between the Protocol and other agreements constitute a somewhat unclear set of intentions. The first point states that parties recognize that "trade and environment agreements should be mutually supportive". This was language proposed by the European Union in lieu of the Miami Group's preferred savings clause. The second point, which was vital to securing the Miami Group's agreement to the Protocol, states categorically that "this Protocol shall not be interpreted as implying a change in the rights and obligations of a Party under any existing international agreements". For the Miami Group, this clause signals a clear and unambiguous intent that the obligations of the WTO will not be undermined by the Protocol. However, the subsequent third point, insisted on by the European Union, states that "the above recital is not intended to subordinate this Protocol to other international agreements" (Cartagena Protocol 2000: Preamble). The unsurprising outcome is that these almost contradictory statements will be interpreted to suit different needs.

With the relationship between the Protocol and other agreements, as well as the scope and extent of legitimate decision criteria (including the precautionary principle), open to different interpretations, it could be concluded that the Protocol, as agreed in 2000, did not categorically shift the advantage to any one side in the ongoing transatlantic GMO conflict

(Safrin 2002; Gupta 2000a, 2000b). However, regardless of surely to-be-continued disputes over precaution and other decision criteria, the European Union sees inclusion of precautionary language in the Protocol as its single achievement. For the European Union, this is perhaps the most concrete benefit from having a global biosafety protocol, since the agreement may help to legitimize its regional approach to GMO regulation, including its potentially precautionary decisions. However, this depends importantly on how the Protocol's provisions will be implemented nationally, and how they will be interpreted and expanded in global forums. It also depends upon developments within the WTO's dispute settlement process and upon the evolving standard-setting activities of the Codex Alimentarius Commission. Global biosafety regime evolution and some new developments are examined next.

3. The evolution of global biosafety governance

This section considers three areas where global biosafety rules are evolving, with implications for regime interplay. These are: negotiation of detailed information-sharing obligations for the agricultural commodity trade within the Protocol (and possible relevance for domestic GMO labelling); global biosafety rule transmission to domestic contexts through capacity-building activities and dispute settlement processes; and evolving memberships in global biosafety regimes.

3.1. Global obligations governing trade in agricultural commodities

A key element in global biosafety regime evolution is the negotiation and implementation of detailed identification and documentation requirements to accompany bulk agricultural commodity shipments that contain genetically modified varieties. As noted above, the information-sharing obligation for these so-called LMO-FFPs (LMOs transferred for food, feed or processing) was one of the most controversial elements during negotiation of the Protocol. The compromise reached called for bulk agricultural commodity shipments containing transgenic varieties to state that they "may contain" LMOs rather than specifying which ones, and for the negotiation of more detailed requirements within two years of the Protocol's entry into force.

Not surprisingly, therefore, elaborating on the "may contain" requirement for the bulk commodity trade was at the centre of conflict at the first Meeting of the Parties to the Cartagena Protocol in Kuala Lumpur in February 2004. Transgenic producer (and exporter) countries, supported by industry, argued that there was no need to go beyond the "may contain" obligation. Others, including the European Union and developing countries, supported by green groups, pushed to agree upon detailed information-sharing obligations, such as the unique identification or genetic transformation code of each LMO variety contained in a particular grain shipment. This proposal was strongly opposed by GMO-exporting countries, but as non-parties to the Protocol they could not block final decisions. The European Union proposal was ultimately blocked by Brazil and Mexico, which are parties to the Protocol, in a sign of new and shifting negotiating alliances. The compromise reached at this meeting deferred agreement on detailed information-sharing obligations to the next Meeting of the Parties (mid-2005 in Montreal) but did "urge" (rather than legally require) parties and others to provide additional information effective immediately (Falkner and Gupta 2004).

Despite efforts by the European Union and most developing countries, particularly the Africa Group, to go beyond this requirement at the 2005 Montreal meeting, no further changes were agreed at this time, primarily because of opposition from New Zealand and Brazil. These countries adopted positions similar to those espoused earlier by the ex-Miami Group on this issue (Earth Negotiations Bulletin 2005). Reaching agreement on detailed documentation requirements for the bulk agricultural commodity trade became thus the main issue for the third Meeting of the Parties in Curitiba, Brazil, in March 2006. While the focus at first was on bringing New Zealand and Brazil closer to an agreement, ultimately it was other countries, particularly Mexico, together with Paraguay, that continued to oppose going beyond the "may contain" obligation (Earth Negotiations Bulletin 2006).

Depending upon who shapes the debate and how far parties can go, the documentation aspect of biosafety treaty evolution is likely to have a substantial impact on the global agricultural commodity trade. This is also linked to domestic initiatives for consumer labelling of GM products. The Protocol does not address domestic labelling, although conflicts over information to accompany commodity shipments are fuelled in part by importing countries' domestic labelling imperatives. The WTO, on the other hand, does have a bearing on domestic labelling for GM foods, although it remains uncertain which WTO agreement is applicable. If the aim of domestic labelling is to provide safety information about impacts on human health, the SPS Agreement would apply. If, however, the aim is to provide information to facilitate consumer right-to-know based on a production process, the TBT Agreement is likely to be applicable, as discussed earlier. It is noteworthy that the issue has not been put to the test globally, insofar as no domestic labelling laws have yet been questioned as incompatible with WTO obligations.

However, future deliberations within the Protocol on more detailed identification requirements for LMO varieties in the agricultural commodity trade may affect this situation. This is particularly the case for the thorny issue of thresholds for identifying genetically modified content in bulk agricultural commodity shipments. Threshold levels identify the percentage content of genetically modified material in a bulk shipment below which information-sharing obligations will not be triggered.[5] Industry groups have argued for a 5 per cent threshold level to be applied across the board for all types of LMO shipments (i.e. the identification requirement stating that the shipment "may contain" LMOs should apply only to shipments containing 5 per cent or more of LMO content). For industry groups, this is a realistic level that would limit disruption to existing commodity trade flows and would guarantee that the adventitious presence of LMOs resulting from cross-pollination or accidental mixing of seed does not require shipments to be identified as containing LMOs. LMO-exporting countries such as Canada and the United States support this industry position. A 5 per cent threshold has also been included in a controversial trilateral agreement between the North American Free Trade Agreement countries of Mexico, the United States and Canada in regulating agricultural commodity trade between them (Gupta and Falkner 2006).

The issue is further complicated by the fact that some parties to the Protocol have already established GMO threshold levels for domestic labelling. The diverse approaches chosen so far are likely to make the search for an international standard more difficult.[6] Moreover, existing national rules differ not only with regard to threshold levels but also with regard to the GM products covered. Countries will thus continue to seek to have their domestic approach supported by global rules. The issue of thresholds and the labelling of GM food is also on the agenda within the Codex Alimentarius Commission. This is likely to remain an important and controversial aspect of global biosafety regime evolution and interplay in the near future.

3.2. *Evolving channels of global rule transmission to a domestic context*

In considering global regime evolution and interplay, it is also useful to examine vehicles of biosafety rule transmission to national contexts, since these can affect the evolution of global regimes. Two channels through which global biosafety rules are being diffused are capacity-building initiatives under the Protocol, and dispute settlement and regime compliance mechanisms, which can provide powerful incentives to abide by global rules and can determine which rules acquire greater validity in both global and national contexts.

3.2.1. Capacity-building initiatives as a vehicle of global rule diffusion

The Protocol calls on countries, international organizations and the private sector to participate in capacity-building to help implement it. These initiatives are now under way and constitute a key ongoing activity in global biosafety governance. One challenge is how to coordinate and ensure compatibility amongst diverse initiatives, each of which may promote distinct approaches to domestic GMO regulation. Capacity-building under the Protocol is intended to be a country-driven process of adopting and adapting biosafety frameworks from elsewhere. Yet, with the process just beginning, it is unclear whether it will spread similar biosafety frameworks across the globe and make trade in GMOs easier (as exporters desire) or whether diverse country-specific biosafety frameworks will emerge that augment importer choice.

The jury is still out, yet this will clearly influence (at least the perceptions of) conflicts between global biosafety regimes. In particular, if membership in the Protocol remains selective, as discussed further below, non-party GMO-exporting countries could use the possibility for capacity-building (including through bilateral agreements and with involvement of the private sector) to target countries that are key potential markets for GMO imports and assist them with establishing biosafety frameworks. Depending upon the regulatory model that is diffused, capacity-building initiatives could either encourage or stymie transfers of GMOs worldwide and contribute to harmonized or diverging national regulatory approaches.

3.2.2. Dispute settlement and compliance mechanisms as vehicles of rule diffusion

The WTO's dispute settlement mechanism (DSM) allows trade disputes to be brought by member states against each other in cases of alleged non-compliance with WTO disciplines. DSM rulings are binding and can be enforced through the use of economic sanctions. This encourages a relatively high degree of compliance with WTO rules among member states, including in the developing world (and certainly it is a stronger enforcement mechanism than many MEAs have). The DSM can thus be a powerful vehicle for the transmission of WTO-compatible procedures to a domestic context when countries choose to abide by its rulings. It is noteworthy, however, that this has not occurred in the most high-profile case of disputed domestic health and safety measures and their impacts on trade – the beef hormone conflict. Although the WTO DSM ruled in that case that the European Union was in contravention of its SPS obligations, the European Union has not altered its regulatory stance, despite the imposition of sanctions (Isaac and Kerr 2003; Kastner and Pawsey 2002).

Yet, this ability to withstand sanctions is often limited to developed countries. The potential for the dispute settlement process to diffuse the trade regime's procedures to domestic contexts was put to the test in the case brought by the United States (joined by Argentina and Canada) against the European Union's regulatory approach towards transgenic crops. The case was widely perceived as an indirect evaluation of the relationship between the Cartagena Protocol on Biosafety and the SPS Agreement (Isaac and Kerr 2003). The final ruling in May 2006 found largely in favour of the complaint by the United States, stating that aspects of the European Union's GMO regulations were operating in contravention of the SPS Agreement. The WTO panel further concluded that it was "not necessary or appropriate" to consider other treaties in reaching its decision (ICTSD 2006). The implications of this ruling for global biosafety regime interplay are important to consider, particularly since the dispute settlement and compliance mechanisms of MEAs, such as the Cartagena Protocol, are often weaker than those of the WTO.

Although the Protocol can avail itself of the CBD's dispute settlement mechanism, it has a separate compliance procedure. This was agreed at the first Meeting of the Parties in Kuala Lumpur, where compliance was a key item on the agenda. Parties can trigger the Protocol's compliance mechanism in cases of their own non-compliance (self-trigger) or if they are "affected or likely to be affected, with respect to another Party" (party-to-party trigger – strongly pushed by the European Union). Measures to facilitate compliance with the Protocol are, so far, the provision of technical and financial assistance, issuing a caution and publishing a case of non-compliance with the Biosafety Clearing-House.

Despite having fought hard for a biosafety protocol, developing countries argued in Kuala Lumpur for a purely facilitative (or "weak") compliance mechanism, instead of one that would also permit punitive sanctions for non-compliance (as does the WTO). Not surprisingly, developing countries were concerned that they might be faced with punitive measures in instances of capacity-related non-compliance. Outspoken opposition to punitive sanctions also came from GMO-exporting countries, but most of these are not yet party to the Protocol and hence were unable to block final decisions. The European Union was the only supporter of punitive sanctions, given its strong interest in an effective protocol as a buffer against potentially unfavourable WTO rulings on its GMO regulations. However, given opposition from developing countries that are parties to the Protocol, the thorny issue of punitive sanctions was deferred to future negotiations.

The possibility of punitive sanctions remains particularly important for the Protocol, given that it does not prohibit trade with non-parties (as do certain other MEAs with trade impacts, such as the Montreal Protocol on Substances that Deplete the Ozone Layer or the Basel Convention on hazardous wastes). Instead of prohibiting trade with non-parties, the Cartagena Protocol in fact permits bilateral agreements between parties and non-parties, as long as such agreements are consistent with its objectives. In such a context, a compliance mechanism that allows for punitive sanctions (including the possibility of trade restrictions) can acquire much greater importance, with possible repercussions for regime interplay as well.

3.3. Membership and participation in global biosafety governance

In contrast to the SPS Agreement (where all WTO members are bound by its obligations),[7] the Cartagena Protocol required at least 50 ratifications to come into force. The Protocol has 141 parties (as of June 2007), yet a few important countries are not members. Most current parties to the Protocol represent the importer rather than the exporter perspective on trade in GMOs. Countries that remain outside the regime include most producers of transgenic crops, such as the United States, Argentina and Canada. A striking feature of the first Meeting of the Parties in Kuala Lumpur in February 2004 was that potential GMO importers, and particularly the European Union, were in the driving seat of treaty evolution. Countries not party to the Protocol no longer had an equal voice in treaty evolution at this meeting, as they had during the initial regime creation. At the same time, a few parties to the Protocol, such as Mexico and Brazil, chose to voice the positions of the ex-Miami Group on key issues such as the documentation requirements for agricultural commodity trade. This trend continued in the second and third Meetings of the Parties in 2005 and 2006, where New Zealand, Mexico and Paraguay, among others, used their position as parties to the Protocol to give voice to ex-Miami Group concerns and block consensus on stringent documentation obligations (Earth Negotiations Bulletin 2005, 2006).

These shifting biosafety-related global alliances have consequences for regime interplay. Brazil is one of the large developing countries with both GMO importer and exporter concerns, while Mexico has to contend with the fact that it is party to a Protocol to which its main agricultural trading partners, the United States and Canada, are not party. Paraguay approved GM soya in 2004 and, according to one source, is now the seventh-largest biotech crop country by hectarage (James 2006). These countries have thus emerged at the meetings of the Protocol as opponents of more stringent documentation obligations for agricultural commodities. Their perspectives signal shifts in alliances across developed and developing countries in future biosafety regime evolution, with implications for how regime interplay will evolve as well. Brazil, Mexico and New Zealand all, for example, joined the United States as observers in the WTO case against the European Union's GMO regulations.

In general, there is currently a three-tier stratification of countries seeking to influence the evolution and implementation of the Protocol and hence of global biosafety rules: countries that (a) are parties to the Protocol; (b) have signed but not ratified the Protocol; and (c) have neither signed nor ratified the Protocol (or the parent CBD, as in the case of the United States). This gives varying degrees of influence to these different groups in regime evolution. Furthermore, it gives rise to a situation where, although rules are developed through some degree of discussion and compromise with all the above groups, the burden of implementation and particularly of financing Protocol obligations falls largely on parties to the agreement. Brazil, in particular, as one of the largest contributors to the Protocol's budget, has voiced this concern. This can influence the effectiveness and future evolution of the regime.

More generally, this party versus non-party negotiating dynamic will be central to how the Protocol develops in the immediate future. It does pose a dilemma for GMO-exporting countries: should they ratify the Protocol in order to participate more fully in treaty evolution in the future? To do so, however, would require immediate compliance with the Protocol's obligations and could also mean being at a competitive disadvantage to agricultural exporting competitors that remain outside the regime (the United States, for example, cannot ratify the Protocol because it has not yet ratified the parent CBD). Not to join, however, is to allow potential GMO-importing countries to shape treaty evolution, with uncertain consequences for GMO trade and access to new markets. It may also give greater influence to GMO importers in the continuing evolution of the Protocol's relationship with WTO agreements. Membership issues will thus be a critical element to monitor in analysing regime dynamics and institutional interplay in the case of biosafety.

4. Conclusion: The relevance of biosafety regime interplay

The above discussion provides a basis from which to consider, as do subsequent chapters in this volume, how the components of an emerging global biosafety framework relate to one another. This can be ascertained through comparing the provisions of global treaties but also from analysing the evolution and interpretation of global rules and how these are influencing each other at the international level. As subsequent analyses in this volume also discuss, concerns over conflicts with the SPS Agreement not only have shaped negotiation of the Protocol but may well have prevented certain countries from ratifying the agreement, affecting its future prospects. At the same time, the coming into force of the Protocol is, in turn, likely to influence the evolution and interpretation of SPS provisions in global forums as well.

The overview of global biosafety rules undertaken here also reveals that, in the face of conflicting transatlantic GMO regulatory approaches, it remains an open question whose regulatory approach to GMOs will become the global standard, if any. The dynamics of regime evolution in the Cartagena Protocol and its interplay with the WTO after the 2006 ruling will continue to clarify which side of the transatlantic divide over GMO regulation will succeed in spreading its approach worldwide.

Ultimately, it is because of the impact on domestic governance that global regime interplay matters. In the case of biotechnology, the stakes are enormous. Developing countries, in particular, have to determine their regulatory choices not simply by considering appropriate technology use in their own context, but also by considering how the transatlantic conflict over GMOs might affect biotechnology use in the future. Thus, empirical understanding of whether and how global regimes are influencing national-level institutional change – i.e. a focus on global–national linkages – is urgently needed. Some analyses are now becoming available of the extent to which domestic regulatory policies are compatible with global biosafety regimes (Baumüller 2003; see also Millstone and Zwanenberg 2003). A recent study of the influence of global biosafety regimes on regulatory policies in Mexico, South Africa and China reveals, for example, that such influence is mediated via domestic institutions and politics and is not uniform or straightforward. In all three cases, however, ratification of the Cartagena Protocol has strengthened the hand of domestic biosafety constituencies vis-à-vis those pushing for a purely trade-facilitative domestic approach to GMOs (Gupta and Falkner 2006).

The diverse analytical approaches to global regime interplay provided by this volume can further assist with such global–national analyses. A research agenda for regime interplay can, moreover, consider whether the global biosafety regimes examined here are bolstered or undermined by proliferating regional and bilateral agreements. This is important also because of the precedent being established for other global trade–environment conflicts. This precedent-setting function is evident from the fact that the Protocol's language on its relationship to other agreements, for example, draws on the compromise reached within the Rotterdam Convention. This borrowing of language highlights the importance of precedent-setting in contentious attempts to govern trade–

environment concerns globally. The relevance of biosafety regime interplay thus transcends the biosafety issue area, making it yet more central.

Acknowledgements

I thank W. Bradnee Chambers, Joy A. Kim and Oran R. Young for comments on an earlier draft and acknowledge institutional support from the Technology and Agrarian Development Group of Wageningen University, where I was based as Visiting Fellow during the preparation of this chapter. This chapter has benefited from a research project on global–national linkages in biosafety governance supported by the John D. and Catherine T. MacArthur Foundation and undertaken jointly with Robert Falkner. I am particularly grateful to Robert Falkner for many useful discussions on these subjects.

Notes

1. The beef hormone conflict was initially governed by the WTO Standards Code, the precursor to the SPS Agreement, which dealt with industrial, health and safety standards for products entering international trade but did not clearly cover process and production methods. This resulted in much controversy over whether the rules of the Code applied to products that differed mainly in their production methods rather than in the end-product. The SPS Agreement was negotiated explicitly to include process and production methods in the oversight of sanitary and phytosanitary measures. Unlike the Standards Code, the SPS Agreement is legally binding on all WTO members. The beef hormone dispute has been the subject of WTO dispute settlement panel and appellate body reports (e.g. WTO 1998). For an early history of the dispute and its initial mediation through the Standards Code, see Halpern (1989). More recent analyses include Biermann (2001) and Kastner and Pawsey (2002).
2. Rotterdam Convention (1998); Basel Convention (1989).
3. Although such figures are hard to come by and even harder to verify, one source claims that the worldwide area devoted to transgenic crops in 2006 was 102 million hectares, and transgenic crops were grown in 22 countries. Of these, 8 countries grew 98 per cent of all transgenics: the United States led with 53 per cent, followed by Argentina (18 per cent), Brazil (11 per cent), Canada (6 per cent), India (3.7 per cent), China (3.4 per cent), Paraguay (2 per cent) and South Africa (1 per cent). Other countries growing transgenic crops in 2006 (in order of hectarage) were Uruguay, the Philippines, Australia, Romania, Mexico, Spain, Colombia, France, Iran, Honduras, the Czech Republic, Portugal, Germany and the Slovak Republic. This suggests that, since late 1999/early 2000 when Protocol negotiations were very intense, more developing countries have started to produce transgenic crops. Countries growing transgenic varieties in 2006 (but not in 1999–2000) include Brazil and India. In 2006, the main transgenic crops grown were soybean (57 per cent of the total global area of transgenics), followed by maize (25 per cent), cotton (13 per cent) and canola (5 per cent). The main genetic modification continued to be herbicide tolerance (68 per cent). In 2006, the global market value of GM crops was estimated

to be US$6.15 billion, representing 16 per cent of the US$38.5 billion global crop protection market and 21 per cent of the global commercial seed market (James 2006).
4. However, the Protocol also makes reference to Principle 15 of the Rio Declaration on Environment and Development (on the precautionary principle) in its Article 1 on objectives. This increases the potential for multiple interpretations of the precautionary actions that are permissible. For example, Principle 15 requires that precautionary measures be cost effective – a criterion not included in the Protocol's Articles 10.6 or 11.8. On the other hand, Principle 15 is also interpreted by critics as too broad and calling for zero risk in its reference to precautionary actions in the face of a "lack of full scientific certainty" about the absence of risk.
5. This discussion of thresholds is taken from Falkner and Gupta (2004).
6. The European Union's GMO labelling and traceability rules, which entered into force in April 2004, require that all food and feed containing GMOs, as well as food produced from or containing ingredients produced from GMOs, be labelled as containing GMOs. In the case of the adventitious presence of GMOs, any product containing more than 0.9 per cent of approved GM material is to be considered a GM product. A 0.5 per cent threshold applies for the adventitious presence of GMOs not yet formally authorized. Australia and New Zealand have adopted a threshold of 1 per cent, South Korea 3 per cent, and Japan and Indonesia 5 per cent. Russia is to lower its 5 per cent threshold to 0.9 per cent and China has recently introduced a 0 per cent threshold for its labelling scheme (Falkner and Gupta 2004).
7. Although participation in the SPS Agreement is high, it is not uniform across developed and developing countries. As of 2001, whereas 92 per cent of high-income countries and 83 per cent of middle-income countries were WTO (and hence SPS) members, only 62 per cent of low-income countries were members. Participation by lower-income countries in the Codex Alimentarius Commission is relatively higher (Hensen and Loader 2001). Participation in global regimes, including those relevant to the WTO, acquires yet more importance in light of a potential trend to adopt global standards by voting, as discussed earlier.

REFERENCES

Appleton, Arthur E. (2000), "The Labeling of GMO Products Pursuant to International Trade Rules", *New York University Environmental Law Journal* 8: 566–578.

Bail, Christoph, Robert Falkner and Helen Marquard, eds (2002), *The Cartagena Protocol on Biosafety: Reconciling Trade in Biotechnology with Environment and Development?* London: RIIA/Earthscan.

Basel Convention (1989), *Basel Convention on the Control of Transboundary Movements of Hazardous Wastes and Their Disposal*, adopted 22 March 1989, entered into force 5 May 1992.

Baumüller, Heike (2003), "Domestic Regulations for Genetically Modified Organisms and Their Compatibility with WTO Rules: Some Key Issues", TKN (Trade Knowledge Network) Paper, International Institute for Sustainable Development, August.

Biermann, Frank (2001), "The Rising Tide of Green Unilateralism in World Trade Law: Options for Reconciling the Emerging North-South Conflict", *Journal of World Trade* 35(3): 421–448.

Boutrif, Ezzeddine (2003), "The New Role of Codex Alimentarius in the Context of the WTO/SPS Agreement", *Food Control* 14: 81–88.
Brack, Duncan, Robert Falkner and Judith Goll (2003), *The Next Trade War? GM Products, the Cartagena Protocol and the WTO*. London: RIIA.
Buckingham, D. E. and P. W. B. Phillips (2001), "Hot Potato, Hot Potato: Regulating Products of Biotechnology by the International Community", *Journal of World Trade* 35(1): 1–31.
Cartagena Protocol (2000), *Cartagena Protocol on Biosafety to the Convention on Biological Diversity: Text and Annexes*. Montreal: Secretariat of the Convention on Biological Diversity.
Christoforou, Theofanis (2000), "Settlement of Science-Based Trade Disputes in the WTO: A Critical Review of the Developing Case Law in the Face of Scientific Uncertainty", *New York University Environmental Law Journal* 8: 622–648.
Codex [Codex Alimentarius Commission] (2003), *Report of the Fourth Session of the Codex Ad Hoc Intergovernmental Task Force on Foods Derived from Biotechnology*, Yokohama, Japan, 11–14 March, ALINORM 03/34.
Codex [Codex Alimentarius Commission] (n.d.), "Statements of Principle Regarding the Role of Science in the Codex Decision-Making Process and the Extent to Which Other Factors Are Taken into Account", in *Codex Alimentarius Commission Procedural Manual 14th Edition*, "Appendix: General Decisions of the Commission".
Coleman, William D. and Melissa Gabler (2002), "Agricultural Biotechnology and Regime Formation: A Constructivist Assessment of the Prospects", *International Studies Quarterly* 46: 481–506.
Convention on Biological Diversity (1992), *Convention on Biological Diversity (with Annexes)*. Concluded at Rio de Janeiro on 5 June 1992. United Nations Treaty Series, Vol. 1760, 1993.
Cosby, Aaron (n.d.), *A Forced Evolution? The Codex Alimentarius Commission, Scientific Uncertainty and the Precautionary Principle*. Canada: International Institute for Sustainable Development.
Covelli, Nick and Viktor Hohots (2003), "The Health Regulation of Biotech Foods under the WTO Agreements", *Journal of International Economic Law* 6(4): 773–795.
Egziabher, Tewolde Berhan G. (1999), "Safety Denied", *Our Planet* (June).
Earth Negotiations Bulletin (2005), "Summary of the First Meeting of the Ad-Hoc Group on Liability and Redress and the Second Meeting of the Parties to the Cartagena Protocol on Biosafety, 25 May–3 June 2005", *Earth Negotiations Bulletin* 9 (320).
Earth Negotiations Bulletin (2006), "Summary of the Third Meeting of the Parties to the Cartagena Protocol on Biosafety, 13–17 March 2006", *Earth Negotiations Bulletin* 9 (351).
Esty, Daniel C. and Damien Geradin (1997), "Market Access, Competitiveness, and Harmonization: Environmental Protection in Regional Trade Agreements", *Harvard Environmental Law Review* 21(2): 265–336.
Falkner, Robert (2000), "Regulating Biotech Trade: The Cartagena Protocol on Biosafety", *International Affairs* 76(2): 299–313.

Falkner, Robert and Aarti Gupta (2004), *Implementing the Cartagena Protocol: Key Challenges*. Sustainable Development Programme Briefing Paper SDP BP 04/04. London: RIIA.

Global Industry Coalition (1999), "Basic Requirements for a Successful Biosafety Protocol", paper circulated at the Sixth Meeting of the Open-Ended *Ad Hoc* Working Group on Biosafety and the Extraordinary Meeting of the Conference of the Parties to the Convention on Biological Diversity, Cartagena, Colombia, February.

Greenpeace (1999), "Greenpeace International's Comments on the Draft Negotiating Text [of the Cartagena Protocol on Biosafety]", prepared for the Sixth Meeting of the Open-Ended *Ad Hoc* Working Group on Biosafety and the Extraordinary Meeting of the Conference of the Parties to the Convention on Biological Diversity, Cartagena, Colombia, February.

Gupta, Aarti (2000a), "Governing Trade in Genetically Modified Organisms: The Cartagena Protocol on Biosafety", *Environment* 42(4): 23–33.

Gupta, Aarti (2000b), "Creating a Global Biosafety Regime", *International Journal of Biotechnology* 2(1–3): 205–230.

Gupta, Aarti (2001), "Advance Informed Agreement: A Shared Basis to Govern Trade in Genetically Modified Organisms?", *Indiana Journal of Global Legal Studies* 9(1): 265–281.

Gupta, Aarti (2004), "When Global Is Local: Negotiating Safe Use of Biotechnology", in Sheila Jasanoff and Marybeth Long-Martello (eds), *Earthly Politics: Local and Global in Environmental Governance*. Cambridge, MA: MIT Press, pp. 127–148.

Gupta, Aarti and Robert Falkner (2006), "The Influence of the Cartagena Protocol on Biosafety: Comparing Mexico, China and South Africa", *Global Environmental Politics* 6(4): 23–55.

Halpern, Adrian Rafael (1989), "The U.S.–EC Hormone-Beef Controversy and the Standards Code: Implications for the Application of Health Regulations to Agricultural Trade", *North Carolina Journal of International Law and Comparative Regulation* 14: 135–155.

Hensen, Spencer and Rupert Loader (2001), "Barriers to Agricultural Exports from Developing Countries: The Role of Sanitary and Phytosanitary Requirements", *World Development* 29(1): 85–102.

ICTSD [International Centre for Trade and Sustainable Development] (2006), "WTO: Biotech Panel Largely Confirms Interim Findings against EU", *Bridges* 10(32), 4 October.

Isaac, Grant E. and William A. Kerr (2003), "Genetically Modified Organisms at the World Trade Organisation: A Harvest of Trouble", *Journal of World Trade* 37(6): 1083–1095.

James, Clive (2006), "Global Status of Commercialized Biotech/GM Crops: 2006", ISAAA Brief No. 35, Ithaca, NY.

Jasanoff, Sheila (1998a), "Contingent Knowledge: Implications for Implementation and Compliance", in Edith Brown Weiss and Harold K. Jacobson (eds), *Engaging Countries: Strengthening Compliance with International Environmental Accords*. Cambridge, MA: MIT Press, pp. 63–87.

Jasanoff, Sheila (1998b), "Harmonization – The Politics of Reasoning Together", in R. Bal and W. Halfman (eds), *The Politics of Chemical Risk*. Dordrecht: Kluwer Academic Publishers, pp. 173–194.

Kastner, J. J. and R. K. Pawsey (2002), "Harmonising Sanitary Measures and Resolving Trade Disputes through the WTO-SPS Framework. Part I: A Case Study of the US–EU Hormone-Treated Beef Dispute", *Food Control* 13: 49–55.

Klevorick, Alvin K. (1996), "The Race to the Bottom in a Federal System: Lessons from the World of Trade Policy", *Yale Law and Policy Review/Yale Journal on Regulation*, Symposium Issue, "Constructing a New Federalism": 177–186.

McConnell, Fiona (1996), *The Biodiversity Convention: A Negotiating History*. Dordrecht: Kluwer Law International.

Marceau, Gabrielle and Joel P. Trachtman (2002), "The Technical Barriers to Trade Agreement, the Sanitary and Phytosanitary Measures Agreement, and the General Agreement on Tariffs and Trade: A Map of the World Trade Organization Law of Domestic Regulation of Goods", *Journal of World Trade* 36(5): 811–881.

Millstone, Eric and Patrick van Zwanenberg (2003), "Food and Agricultural Biotechnology Policy: How Much Autonomy Can Developing Countries Exercise?", *Development Policy Review* 21(5): 655–667.

Phillips, W. B. P. and William A. Kerr (2000), "The WTO versus the Biosafety Protocol for Trade in Genetically Modified Organisms", *Journal of World Trade* 34(4): 63–75.

Rajan, Mukund Govind (1997), *Global Environmental Politics: India and the North–South Politics of Global Environmental Issues*. Delhi: Oxford University Press.

Revesz, Richard L. (1992), "Rehabilitating Interstate Competition: Rethinking the Race-to-the-Bottom Rationale for Federal Environmental Regulation", *New York University Law Review* 67: 1210–1254.

Rotterdam Convention (1998), *Final Act of the Conference of Plenipotentiaries on the Convention on the Prior Informed Consent Procedure for Certain Hazardous Chemicals and Pesticides in International Trade*, adopted 10 September 1998, UN Doc. UNEP/FAO/PIC/Conf/5 of 17 September.

Safrin, Sabrina (2002), "Treaties in Collision? The Biosafety Protocol and the World Trade Organization Agreements", *American Journal of International Law* 96(3): 606–628.

Sarewitz, Daniel (2000), "Science and Environmental Policy: An Excess of Objectivity", in Robert Frodemen (ed.), *Earth Matters: The Earth Sciences, Philosophy and the Claims of Community*. Englewood Cliffs, NJ: Prentice-Hall, pp. 79–98.

Skogstad, Grace (2001), "Internationalization, Democracy and Food Safety Measures: The (Il)Legitimacy of Consumer Preferences?", *Global Governance* 7: 293–316.

SPS Agreement (1994), *Agreement on the Application of Sanitary and Phytosanitary Measures. Annex 1A to the Final Act Embodying the Results of the Uruguay Round of Multilateral Trade Negotiations*, Marrakesh, 15 April 1994.

TBT Agreement (1994), *Agreement on Technical Barriers to Trade. Annex to the Final Act Embodying the Results of the Uruguay Round of Multilateral Trade Negotiations*, Marrakesh, 15 April 1994.

Third World Network (2000), *Open Letter to Delegates to the Resumed Session of the First Extraordinary Meeting of the Conference of the Parties to the Convention on Biological Diversity*, Montreal, January. Penang, Malaysia: Third World Network.

WTO [World Trade Organization] (1998), "Report of the Appellate Body: EC Measures Concerning Meat and Meat Products (Hormones)", WT/DS26/AB/R, WT/DS48/AB/R, AB-1997-4, 16 January.

WTO (2003), "European Communities: Measures Affecting the Approval and Marketing of Biotech Products: Request for Establishment of a Panel by the United States", WT/DS291/23, 8 August.

Part II
Institutional interplay and its application to biosafety and trade

3

Analysing biosafety and trade through the lens of institutional interplay

Heike Schroeder

1. Introduction

The practice of genetically modifying animals and plants through domestication and controlled breeding is nothing new, but has evolved over many centuries. Modern techniques of genetic modification constitute a significant novelty, and our ability to transfer genes from one species to another, non-related species gives rise to a great number of new possibilities and risks. Modern biotechnology has the potential significantly to increase the production of food around the world to secure food supply for a growing world population and to extend farming to areas where conditions would otherwise be too harsh. Modern biotechnology might be able to halt deforestation and provide a side benefit in terms of reducing global carbon emissions. By stopping the destruction of indigenous forests, it would also reduce the loss of biodiversity. In addition, modern biotechnology creates new business opportunities and markets and is a potentially rich source of revenue for countries exporting agricultural produce.

In terms of risks, modern biotechnology has already been shown to affect the ecosystem adversely. Transgenic plants transmit their genes to other crops or wild plants through pollen dispersal. It is conceivable that they may evolve into invasive species if their superior traits allow them to out-compete other plants. Modern biotechnology also gives rise to concern about long-term health impacts that are barely understood. Genetically modified food may trigger new allergies, possibly at unprecedented

Institutional interplay: Biosafety and trade, Young, Chambers, Kim and ten Have (eds), United Nations University Press, 2008, ISBN 978-92-808-1148-3

levels. Small-scale farming systems could be negatively affected, and the production of mono-crops, with all its shortcomings, might be encouraged. Even a security risk is perceivable if genetic engineering were to produce highly contagious microbes for which a cure had yet to be developed (Safrin 2002). In addition, the patenting of seed by Monsanto prevents small farmers from saving seeds to plant, thus tying them to perpetual purchase of Monsanto seed and the necessary herbicides and pesticides. This dependency becomes even more dangerous in light of studies that show that biotechnically modified crop yields are much lower than promised or advertised (Shiva 1998).

Balancing these potential benefits and risks of biotechnology is a major challenge, and is unfolding in particular in the realm of international trade rules, specifically the agreements under the World Trade Organization (WTO) and the Cartagena Protocol on Biosafety (Biosafety Protocol) to the Convention on Biological Diversity (CBD). The Biosafety Protocol, which was adopted in 2000 and came into force in 2003, seeks to establish practical rules and procedures for the safe transfer, handling and use of genetically modified organisms (GMOs). This relatively new regime is faced with the challenge of evolving in a manner acceptable to both GMO-exporting countries, which are eager to reap the benefits of modern biotechnology, and GMO-importing countries, which seek to avoid the inherent risks. The majority of the current parties to the Biosafety Protocol tend to represent the importer, rather than the exporter, perspective on GMO trade. The exporter perspective is represented more persuasively by the WTO's Agreement on the Application of Sanitary and Phytosanitary Measures (SPS Agreement), which sets out a range of requirements designed to avoid protectionist misuse of health and environmental regulations. Because these two agreements may apply concurrently, the interplay between them has been the subject of much discussion (see, for example, Anderson 2002; Safrin 2002; Brack et al. 2003; Isaac and Kerr 2003; Winham 2003). This chapter examines this interplay, drawing on the insights developed as part of the project on the Institutional Dimensions of Global Environmental Change (IDGEC).

2. The Institutional Dimensions of Global Environmental Change (IDGEC) Project

The IDGEC project,[1] launched in 1998 and concluded in 2007, was a core research project of the International Human Dimensions Programme on Global Environmental Change (IHDP).[2] The IDGEC Science Plan (Young et al. 1999, revised 2005) charted the project's research agenda, which sought to understand what roles institutions play as determinants of the course of human–environment interactions, focusing especially on global environmental change. The project defined institutions as "clusters of rights, rules, and decision-making procedures that give rise to social practices, assign roles to participants in these practices, and govern interactions among occupants of these roles". Hence, they represent social artefacts and make up the "rules of the game" within which the players operate. Organizations, in turn, refer to organizational entities that typically have a budget, a location and personnel (Young 2002a).

The objective of increased understanding of the role of institutions directed IDGEC's three research foci: *causality*, investigating the roles that institutions play in causing and addressing global environmental change processes; *performance*, looking into why some institutional responses to global environmental change appear to be more successful than others; and *design*, assessing the effectiveness of various existing designs and exploring the prospects for designing new institutions to cope with specific cases of environmental change. These foci can be explored through an examination of the respective degrees of fit, dimensions of scale and dynamics of interplay, as explained below (Young 1999a, 2001; Underdal 2002; Underdal and Young 2004).

IDGEC's analytic themes deal with factors that both determine and are determined by the causality, performance and design of institutions governing human–environment relations. They include the following:

- The *problem of fit* probes whether prevailing institutional arrangements are well enough matched to the properties of the biophysical system to which they relate. Lately, several researchers have extended the notion of fit to include the degree of match with socio-economic systems as well (Ebbin 2002; Young 2002b).
- The *problem of interplay* scrutinizes whether distinct institutional arrangements interact with others horizontally or vertically and politically or functionally in ways that significantly influence outcomes (Young 1996, 2002c; Lebel 2005).
- Lastly, the *problem of scale* considers to what extent findings about the roles institutions play can be generalized across levels in spatial, temporal and jurisdictional scales (Young 1994; Alcock 2002).

IDGEC has generated knowledge regarding these research foci and analytic themes by carrying out flagship activities that are linked to theoretical concerns and also provide empirical domains. They cover *terrestrial systems* with research on the Political Economy of Tropical and Boreal Forests (PEF), *atmospheric systems* with a Carbon Management Research Activity (CMRA) and *oceanic systems* with research on the Performance of Exclusive Economic Zones (PEEZ) (Hoel et al. 2000; Sewell et al. 2000; Contreras et al. 2001; Young 2003a). Figure 3.1 depicts the various components of IDGEC's research agenda.

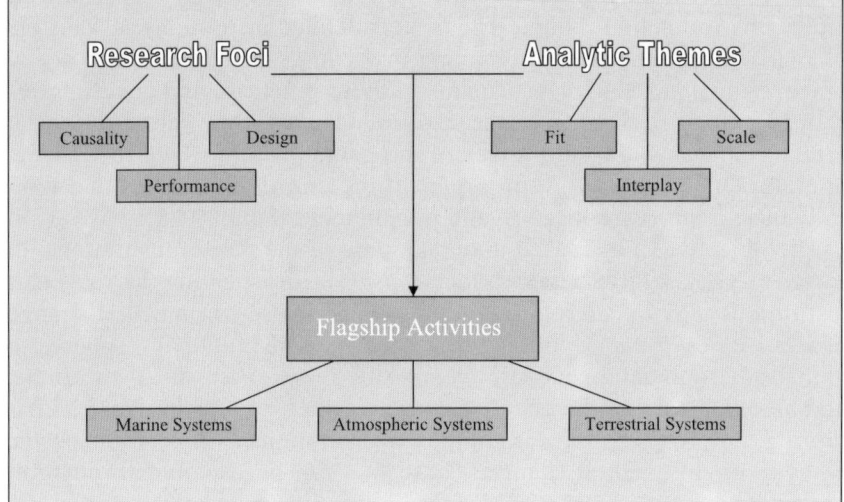

Figure 3.1 IDGEC's research agenda
Source: Institutional Dimensions of Global Environmental Change, *Biennial Report: Spring 2001–Spring 2003*, Santa Barbara, CA, p. 6.

3. IDGEC's typology of interplay

Interplay occurs when at least one institution affects the performance of at least one other institution by evoking a change in behaviour among the players governed by this institution. Institutions operate at all levels of social organization – at local, national and international levels – and can be small or large in size and narrow or broad in scope. This section examines the performance–interplay relationship, directions and types of interaction, forms of dependence and categories of structural linkage.

3.1. The performance–interplay relationship

Many analytical distinctions and criteria for evaluating the performance or effectiveness of institutions exist (Young 1999b, 2001, 2003b; Helm and Sprinz 2000; Miles et al. 2002; Underdal 2002; Underdal and Young 2004). Criteria can be economic (efficiency, cost-effectiveness), political (equity) and ecological (sustainability). The variables selected can be endogenous, and may account for variations in the effectiveness of institutions by influencing the character of the institutional arrangements themselves. Or they can be exogenous, and include the physical, biological and social conditions that make up the environment in which the institution operates. Variables can also include the linkage or fit between the institutional characteristics of a governance system and the environment in which it is expected to function (Young 1999a).

A widely used approach to evaluating performance differentiates between *output* (rules and norms), *outcome* (behavioural change) and *impact* (change in collective outcome or final effect) levels. Interplay occurs at the *output* level when a change in one institutional setting, i.e. the "source" institution (Oberthür and Gehring, Chapter 5 in this volume), affects the interpretation of the rules or leads to ambiguity about how rules should apply in another institutional setting, i.e. the "target" institution. Interplay occurs at the *outcome* level when a change in the "source" institution as the independent variable affects the performance of another institution (the "target" institution or dependent variable) through a change in the behaviour of players governed by the "source" institution. Lastly, interplay occurs at the *impact* level when institutional change at the "source" institution triggers a shift in the scope of the problem dealt with by the "target" institution.

3.2. Directions and types of interaction

Institutions interact both *horizontally*, i.e. at the same level of social organization, such as between the biodiversity and climate regimes (Kim 2004), and *vertically*, i.e. across levels of social organization, such as between the Biosafety Protocol and national implementing agencies (Gupta and Falkner 2006).

Interplay between institutions can be either *functional* or *political* in nature. Functional interplay occurs when a problem addressed by multiple institutions is linked in biophysical or socio-economic terms, meaning that it stems from a systemic interdependence among the related institutions. An example of a biophysical linkage leading to functional interplay occurs between the Montreal and Kyoto Protocols as both treaties seek to regulate substances that are both ozone depleting and a greenhouse gas. Substitutes for CFCs, such as HCFCs and HFCs, while being much less harmful to the ozone layer, may have the unfortunate side effect of global warming at significantly higher potency than that of carbon dioxide (Oberthür 2001). Another example is interplay between the WTO trade rules and the Kyoto climate regime (Kim 2001; Stokke 2004). Socio-economic linkages leading to functional interplay can be found whenever there is competition for a scarce resource or a budgetary constraint, such as among ministries in competition for limited government funds. Or they can occur when a socio-economic activity such as trade in goods or services is regulated by different authorities, such as between the North American Free Trade Agreement (NAFTA) and the Convention on International Trade in Endangered Species (CITES). While CITES establishes lists of endangered species that are prohibited from being traded internationally, NAFTA seeks to reduce trade restrictions among its member countries (Vogel 1997).

Political interplay, on the other hand, arises when players seek to link (or de-link) institutions intentionally or engage in institutional bargaining in the interest of pursuing certain objectives. Interplay thus arises from the consequences of institutional design (Young 2002a). This intended interplay is then likely to result in behavioural changes in the players governed by these institutions. But what are the driving forces behind political interplay? I argue that there are three main types of such interplay.

The first type, called *issue-based political interplay*, proposes that interplay is derived from the motivation of players to enhance institutional effectiveness in solving a given problem. Issue-based political interplay usually stems from a situation where the problem at hand is sufficiently pressing and affected significantly by functional interplay with another institution. Players will choose the most *effective* option to find an agreeable solution. An example here would be the creation of the Forest Stewardship Council in 1993 as a result of the Statement of Forest Principles adopted at the 1992 Rio Summit not going far enough to protect the world's remaining forests.

The second type, *goal-based political interplay*, maintains that interplay arises from the intent of players to improve economies of scale or efficiency. This often happens by establishing some method of collaboration between or among institutions. Goal-based interplay would occur, for instance, if two regional emissions trading systems were to hook up to one another to increase the size of their market in order to reduce the price of carbon. Here, the more attractive option is whichever is most *efficient*, even if it were to come at the cost of effectiveness of impact.

The third type is termed *power-based political interplay*, indicating that interplay in this case is motivated by the intent of players to increase their bargaining power or influence. Interplay is likely to arise when at least one dominant actor or group of actors seeks to profit from a change in behaviour among the players involved or from an explicit or implicit institutional bargaining agreement. One example of power-based interplay is the Russian ratification of the Kyoto Protocol to win the European Union's support for its entry into the WTO. Another is Japan's bargaining for a more generous carbon sequestration allowance to make it easier for Japan to reach its Kyoto emissions reduction target in exchange for its agreement to ratify the Kyoto Protocol (and contribute to enabling the Kyoto Protocol to come into effect) and its effect on Japan's domestic forestry sector (Schroeder 2003).

3.3. Forms of dependence

Dependence between or among institutions can take either *reciprocal* or *unidirectional* forms. Usually it occurs somewhere in between these polar categories, and is, to some extent, asymmetrical, one institution exerting somewhat more influence over the other (Young 1996). Reciprocal interaction, for example, is a common feature between or among economic and regulatory arrangements, such as between NAFTA (free trade) and CITES (international trade in wildlife).

Unidirectional linkages, on the other hand, arise when the operation of one institution affects others significantly without triggering notable responses. Vertical interplay often tends to trigger such unidirectional interaction. For example, the European Union's ban on importation of certain seal products has significantly impacted the livelihoods of local communities that are dependent on trade in seal products. But these communities lack the institutional capacity to respond politically and effectively to such institutional change at the level of the European Union (Young 2002a: 84). The more reciprocal the interaction, the greater the incentive is likely to be for all affected parties to work toward a commonly beneficial institutional adjustment if the interaction generates negative side effects.

3.4. Categories of structural linkage

Institutions can be structurally linked in several ways. Young (1996) distinguishes between embedded, nested, clustered and overlapping institutions.

3.4.1. Embedded institutions

Embedded institutions are institutions that are deeply embedded in overarching institutional arrangements operating under implicit broader principles and practices. Most issue-specific regimes in international society are embedded institutions that are based on an understanding of international society and its rules of state sovereignty (King 1997). The 1973 Agreement on the Conservation of Polar Bears, components of the Antarctic Treaty System and the emerging environmental protection regime for the Arctic are examples of institutional embeddedness (Young 1996). The international economic and financial order established by the 1944 Bretton Woods system and the international financial institutions (World Bank Group, International Monetary Fund) operating under it represent another such example. The basic assumption informing embedded institutions here is that a market free of barriers to the flow of private trade and capital will generate maximum economic wealth.

3.4.2. Nested institutions

Nested institutions are described as linkages where specific arrangements restricted in terms of functional scope or geographic domain are nested in broader institutional frameworks that deal with the same general issue area, but that are less detailed in terms of their application to specific problems. A common type of institutional nesting is that of substantive protocols into agreements established by framework conventions taking the so-called convention-protocol approach (Susskind 1993). Examples of nesting include the Vienna Convention on ozone-depleting substances, its Montreal Protocol and subsequent agreements to accelerate and broaden the phase-out of ozone-depleting substances (Benedick 1998); the integration of the newer sulfur dioxide, nitrogen oxide and volatile organic compounds protocols into the older framework established by the 1979 Convention on Long-Range Transboundary Air Pollution (LRTAP) is another instance of institutional nesting. Institutional nesting gives a regime the capacity to effectively adapt to changes that may occur, such as the advancement of scientific knowledge or alterations in the biophysical properties of the environmental problem.

3.4.3. Clustered institutions

Clustered institutions arise in a situation where several institutional arrangements are gathered into an institutional framework or package. This is most likely to be the result of political bargaining or increasing economies of scale in the operation of the regime, where the end result is likely to constitute a net benefit to all participants. An example of institutional clustering is the 1982 United Nations Convention on the Law of the Sea, which incorporates functionally different provisions for navigation, fishing, deep seabed mining, marine pollution, scientific research, etc. What holds these different elements together is a common concern for issues relating to the marine environment. An example where an institutional cluster has not yet been created despite some functional overlap is the atmosphere, where a law of the atmosphere would cluster the regimes for transboundary air pollution, ozone depletion, and climate change (Young 1996).

3.4.4. Overlapping institutions

Lastly, overlapping institutions refer to individual regimes that were formed for different purposes and largely without reference to one another, but whose policy goals and regulations intersect or overlap. Such regimes would be addressing a common issue or problem with different policy objectives. This may lead them to have substantial impacts on one another. One common example of interplay between overlapping institutions is seen in trade and environment regimes, where the environment may be positively or negatively impacted by free trade policies, or where free trade may be fostered or frustrated by environmental policies (King 1997). Overlap can, in other words, generate positive or negative results; it can produce either opportunities for coordination between or among institutions to derive synergies, or it can trigger conflict from diverging rules and norms that may produce stumbling blocks for either all institutions involved (reciprocal interaction) or only some or one (unidirectional interaction). According to Young (1996), it may be possible in some cases to avoid the unanticipated negative results of overlapping institutions by creating institutional clusters that are more likely to foster approaches amenable to managing this interplay.

These relationships can be summarized in a matrix of the typologies of interplay (see Table 3.1). This gives examples for horizontal, vertical, functional and political interplay, indicates their levels of reciprocity by aligning the examples left (reciprocal), centre (somewhat reciprocal) or right (unidirectional) in their respective boxes, and specifies into which category of structural linkage these examples fall.

4. Interplay between the biosafety and trade regimes

This section applies this typology to the case of interplay between the biosafety and trade regimes. The objective is to improve understanding of the underlying driving forces and potential consequences of interplay in the biosafety and trade arenas, and to determine the impacts of such interplay on the performance of the institutions involved.

4.1. Determining the direction and type of interaction

The direction of interaction between the biosafety and trade regimes is *horizontal*, given that both operate at the international level. The two regimes exhibit a *functional linkage* because both define governance approaches on trade in transboundary GMOs. Whereas the Biosafety Protocol regulates trade solely in "living modified organisms" (LMOs),[4] the WTO seeks to liberalize international trade in general, thus implicitly seeking to liberalize trade in transboundary LMOs. The functional linkage is therefore a *socio-economic* one.

Because of this functional link, the type of *political interplay* to be expected to occur in order to reconcile differences in the two regimes' approaches to trade in GMOs is likely to be *issue based*. Interplay is thus likely to be motivated by the intention of players to enhance institutional *effectiveness* in solving pending questions, such as the relative roles of

Table 3.1 Institutional interplay: Categories of structural linkage (embeddedness, nesting, cluster or overlap)

		HORIZONTAL	VERTICAL
		Reciprocal ↔ unidirectional	Reciprocal ↔ **unidirectional**
FUNCTIONAL (unintended)	Biophysical	Montreal Protocol and Kyoto Protocol on governance of ozone-depleting substances that are also greenhouse gases *Institutional cluster*	Kyoto Protocol and national agencies that play a role in the national implementation process *Institutional overlap*
	Socio-economic	Competition among the ministries of a government over budget allocations *Institutional embeddedness*	Allocation of tax revenues between local and national governments *Institutional embeddedness*
	Issue based (effectiveness)	UN Framework Convention on Climate Change and its Kyoto Protocol *Institutional nesting*	Forest Stewardship Council and national forestry practices *Institutional overlap*
POLITICAL (intended)	Goal based (efficiency)	Chicago Climate Exchange and California Climate Action Registry *Institutional cluster*	EU Emissions Trading Scheme and domestic trading schemes such as the Chicago Climate Exchange *Institutional cluster*
	Power based (profit)	Russian ratification of the Kyoto Protocol to win EU support for Russia's entry into the WTO *Institutional overlap*	Japanese bargaining for higher allowances of carbon sequestration for its support of the Kyoto Protocol and its impact on Japan's forestry sector *Institutional cluster*

precaution versus scientific knowledge or compatible modalities of risk assessment.

4.2. Comparing the performance–interplay relationship and regime principles and rules

As mentioned earlier, interplay can affect the output, outcome and impact levels of institutional performance. Given the young age of the Biosafety Protocol, this section will focus on the output level and look at the interplay between the trade and biosafety regimes at the level of principles and rules.

The general principles under which the two regimes operate differ significantly. This difference centres on the role of science in decision-making under conditions of uncertainty, and on whether importing or exporting countries should bear the burden of providing evidence of risk or safety.

The Biosafety Protocol's underlying principle for achieving its objective is the precautionary approach.[3] The negotiation and adoption of the Biosafety Protocol itself resulted in a shift in discourse and subsequent universal acceptance of the precautionary principle (Andrée 2005; Schroeder et al. forthcoming). The Protocol stipulates in Articles 10.6 and 11.8 that "lack of scientific certainty due to insufficient relevant scientific information and knowledge ... shall not prevent that Party from taking a decision" regarding the importation of living modified organisms. The preamble of the Biosafety Protocol reaffirms the precautionary approach contained in Principle 15 of the Rio Declaration on Environment and Development.

The WTO, by contrast, maintains that trade restrictions are permissible only when they are based on scientific principles. Under the SPS Agreement, these may not be maintained without sufficient scientific evidence, except where relevant scientific evidence is insufficient. In such a case, a WTO member may provisionally adopt sanitary or phytosanitary measures on the basis of available pertinent information (Article 5.7). The SPS Agreement's preamble stipulates that WTO members should not be prevented from adopting and enforcing measures necessary to protect human, animal or plant life or health as long as these measures do not constitute arbitrary or unjustifiable discrimination.

The principles underpinning these regimes differ in the emphasis they give to the role of science in decision-making, and the margin of discretion they concede to policy makers when regulating GMOs (including to ban trade) under conditions of uncertainty. The resulting tension was evident during the negotiation of the Biosafety Protocol, where GMO-exporting countries sought to establish the SPS Agreement's general provisions as the benchmark, whereas GMO-importing countries sought to develop more specific and stricter standards within the Protocol.

The views of these competing camps found expression in the language of the Biosafety Protocol's preamble, which emphasizes the "mutual supportiveness" of trade and environment agreements. It declares that the Protocol should not be interpreted as implying a change in rights and obligations under other agreements (including WTO rules), but also that the Protocol is not subordinated to other agreements (including WTO rules). This mixed language reflects an attempt by the negotiating countries to preclude a potential trade conflict by leaving enough flexibility in interpretation to suit different needs (Gupta 2000). The precise implication of this preamble text is somewhat unclear, and the relationship between these agreements will likely depend on how their specific rules are interpreted and applied in practice. Table 3.2 summarizes the differences in the principles enshrined in the Biosafety Protocol and the SPS Agreement. These principles are reflected in the two regimes' more specific rules and obligations. Issues of interplay may arise in cases where these rules apply concurrently, for example in risk assessment and when responding to scientific uncertainty.

Table 3.2 Principles of regimes governing transboundary movement in GMOs

Regime	Principles: science vs. precaution
Biosafety Protocol	Reaffirms the precautionary approach contained in the Rio Declaration (Preamble)
	The precautionary approach is the underlying principle for reaching the objective of the Protocol (Article 1)
	"Lack of scientific certainty due to insufficient relevant scientific information and knowledge ... shall not prevent that Party from taking a decision, as appropriate, with regard to the import of the living modified organism in question" (Article 10.8)
SPS Agreement	WTO members should not be prevented from adopting or enforcing measures necessary to protect human, animal or plant life or health as long as these measures do not constitute arbitrary or unjustifiable discrimination (Preamble)
	Trade restrictions may not be maintained without sufficient scientific evidence, except where relevant scientific evidence is insufficient. In such a case, a WTO member "may provisionally adopt sanitary or phytosanitary measures on the basis of available pertinent information" (Article 5.7)

In relation to risk assessment, the Biosafety Protocol and the SPS Agreement differ slightly in their approaches to determining and responding to risk. In the Biosafety Protocol, the advance informed agreement procedure requires that an exporter seek the consent of the importing party prior to the first shipment of an LMO destined for intentional introduction into that party's environment (Articles 7 and 8). The importing country has the opportunity to assess risks associated with the LMO before agreeing to or refusing its importation. This decision has to be based on risk assessment (Article 15) (Falkner and Gupta 2004). The two agreements also differ in terms of the nature of the risk assessment: the risk assessment in the Biosafety Protocol allows countries to take account of socio-economic effects as well.

The SPS Agreement requires WTO members to base their measures on an assessment of risks to human, animal or plant life or health, in order to ensure that these measures are not a disguised restriction on international trade and to minimize any negative trade effects. According to the WTO Appellate Body, WTO members must ensure a "rational relationship" between the risk assessment and any measure developed to regulate trade (Appellate Body Report 1998: para. 193). Furthermore, members are expected then to manage risks according to an "acceptable level of risk deemed appropriate by the member state" (para. 177).

This gives rise to two questions. The first concerns the appropriate nexus between the risk assessment and the regulatory measures. Will the SPS Agreement's requirements on risk assessments be applied in a manner consistent with the Biosafety Protocol's detailed requirements relating to risk assessment for transboundary movement of LMOs? The second question addresses the issue of who is responsible for covering the cost of risk assessments. The Biosafety Protocol explicitly states that "the cost of risk assessment shall be borne by the notifier if the Party of import so requires" (Article 15.3). The SPS Agreement, by contrast, establishes no such requirement but states that, where relevant scientific information is insufficient, "Members shall seek to obtain the additional information necessary for a more objective assessment of risk" (Article 5.7). The answer to these questions of interplay depends largely on how the two agreements are interpreted by parties or the adjudicating body in the event of a dispute.

Questions also arise in the case of insufficient scientific information. How should countries respond where there is significant scientific uncertainty? Should uncertainty favour exporting countries (which do not want environmental measures to be used as disguised trade barriers) or importing countries (which, faced with uncertainty, would often prefer not to import GMOs)? Differences in the two regimes may prove particularly important when applied in the politically contentious context of trade in commercially valuable agricultural products destined for consumption as food or as feed or for processing.

These questions are best explored in light of the specific rules of the Protocol and the SPS Agreement. The Protocol, on the one hand, allows parties to take precautionary action if scientific uncertainty prevails (Article 11.8). It states generally that parties "may take a decision on the import of living modified organisms intended for direct use as food or feed, or for processing, under its domestic regulatory framework" (Article 11.4). This right to take decisions is supported by the precautionary approach, mentioned above, which provides that "lack of scientific certainty due to insufficient relevant scientific information and knowledge ... shall not prevent that Party from taking a decision, as appropriate, with regard to the import of that living modified organism" (Article 11.8).

The SPS Agreement, on the other hand, does not explicitly address the precautionary principle, but allows precautionary measures under certain conditions. Article 5.7 of the SPS Agreement stresses that countries have the right to take measures where scientific information is insufficient. In these circumstances, however, they must base their measures on available pertinent scientific information, carry out risk assessment, which must find evidence of an ascertainable risk, and base measures on international standards. The article thus allows for precautionary decision-making only on a provisional basis, until further scientific evidence of harm or lack thereof has been obtained.

The WTO has a functioning compliance mechanism, whereas the Biosafety Protocol's compliance procedures are still developing – they were set up at the first Meeting of the Parties (COP/MOP I) in Kuala Lumpur in April 2004. Under the Protocol, measures against non-compliant parties are restricted to soft-law measures, including giving advice or assisting the party by developing a compliance action plan with a time frame, asking the party in question to submit progress reports, and reporting to the MOP on efforts made by the party. The Protocol does not prohibit trade with non-parties – unlike the Montreal Protocol on ozone depletion and the Basel Convention on trade in hazardous wastes – but permits bilateral agreements between parties and non-parties as long as trade is consistent with the provisions of the Biosafety Protocol (Falkner and Gupta 2004: 9). A party submitting a case to the WTO dispute settlement body may simultaneously submit the same case to the compliance body of the Biosafety Protocol. Such dual submission may put pressure on the WTO and on the Biosafety Protocol to consider likely outcomes under one another's compliance scheme, which may lead to some sort of interjudicial exchange between the two bodies to arrive at a mutually compatible outcome (Oberthür and Gehring, Chapter 5 this volume).

4.3. Comparing the form of dependence and regime memberships

Given the broad overlap in membership between the two regimes, interplay is likely to be largely *reciprocal* but also likely to affect the regime with the smaller membership comparatively more. Some significant countries (in terms of their volumes of GMO exports) are not parties to the Biosafety Protocol, which could generate tension among the two institutions. Whereas the WTO has a membership of 149 countries and the Codex Alimentarius Commission (which is important in this context) involves some 171 members, the Biosafety Protocol has so far been ratified by 141 countries and the European Union; the main GMO-exporting countries – the United States, Canada, Australia and Argentina – have yet to ratify the agreement. GMO-importing countries, such as EU member states and many developing countries, have, in turn, ratified the Protocol. When a GMO-exporting country is not a member of the Biosafety Protocol, but importing and exporting countries are both members of the WTO, a conflict is most likely to be resolved under WTO rules not Protocol rules. This situation could potentially undermine the rights of a GMO-importing party when in conflict with a GMO-exporting non-party to the Biosafety Protocol. Although one can assume that parties to a prior agreement, such as the WTO, would see to it that their commitments under a new, issue-related agreement such as the Biosafety Protocol would not be in conflict with prior commitments, the biosafety and trade regimes still constitute a case of possible tension over their prevalence in the event of a conflict.

4.4. Comparing structural linkages and regime objectives

The biosafety and trade regimes represent a case of *overlapping* institutions that have developed, or are in the process of developing, their own regulatory measures or standards and procedures on transboundary movement in GMOs. The objectives of the biosafety-related institutions differ fundamentally.

The WTO, which was established in 1994 to supersede the General Agreement on Tariffs and Trade of 1947, is very much embedded in the post–World War II international economic order based on the pursuit of liberalizing international trade. The objective of the WTO is thus to remove barriers to the international movement of goods and services. To prevent countries from setting health standards or labelling requirements as disguised barriers to trade, the WTO established the SPS Agreement, to ensure that measures taken by member states to protect human, animal and plant health do not constitute protectionist barriers to trade, and the Agreement on Technical Barriers to Trade (the TBT Agreement), to restrain member states from adopting non-tariff barriers in the form of technical regulations including packaging, marking, and labelling requirements; the SPS and TBT agreements were thus nested into the WTO's institutional structure.

The Codex Alimentarius Commission pursues two main goals: to protect consumer health; and to promote fair practices in food trade. It does this by providing standards, guidelines and recommendations for the safe trading of agricultural commodities. Although the Codex Alimentarius was established as early as 1963, it gained prominence only after the WTO adopted Codex standards as its benchmark in resolving disputes between nations related to trade in food commodities in 1994 – an example of institutional *clustering*. The Codex Alimentarius now plays the important, if challenging, role of providing legitimized food safety standards for WTO member states, because measures taken by its members to protect the health of humans, animals and plants under its provisions are based on Codex standards (Post 2005).

Lastly, the Biosafety Protocol, adopted in 2000 and entered into force in 2003, aims to protect a nation's domestic environment from the release of imported LMOs that have not gone through domestic risk assessment and are not authorized for import, because they could have adverse effects on that country's conservation and sustainable use of biological diversity or pose risks to human health. Thus, whereas the general objective of the Biosafety Protocol is to *regulate* trade to protect GMO-importing countries from adverse effects on their ecosystems, that of the WTO is to *liberalize* trade. The WTO's SPS and TBT Agreements seek to restrain countries from adopting unjustified trade-restrictive measures that might be disguised acts of protectionism. These objectives are not irreconcilable, but may give rise to tension in their application between GMO-exporting countries (including the United States, Canada, Australia and Argentina) and GMO-importing countries (including EU countries). Whereas importing countries often seek to exercise precaution, for example by restricting transboundary movement when the long-term consequences of GMOs on the ecosystem are not yet fully understood, exporting countries tend to seek access to foreign markets by arguing that any trade restrictions should be based on available pertinent science. Hence the Biosafety Protocol was formed for a different objective than the WTO and its agreements, and represents a case of overlapping institutions.

The different objectives of these institutions thus relate to their structural linkages with one another. These different objectives are summarized in Table 3.3.

Table 3.3 Objectives of regimes governing transboundary movement in GMOs

Regime	Regime objectives and structural linkage types			
WTO	To provide a common institutional framework for the conduct of trade relations among its members and expand the production of and trade in goods and services (Article II)	Institutional nesting	Institutional cluster	Institutional overlap
SPS Agreement	To ensure that SPS measures do not represent unnecessary, arbitrary, scientifically unjustifiable or disguised restrictions on international trade, while recognizing the sovereign right of member states to provide the level of health protection they deem appropriate (Preamble)			
TBT Agreement	To ensure that technical regulations and standards, including packaging, marking and labelling requirements, and procedures for assessment of conformity with technical regulations and standards do not create unnecessary obstacles to international trade (Preamble)			
Codex Alimentarius	To develop international standards and standard-setting guidelines for food safety to protect the health of consumers, to ensure fair trade practices in the food trade and to promote the coordination of all food standards work undertaken by international governmental and non-governmental organizations			
Biosafety Protocol	To ensure an adequate level of protection in the safe transfer, handling and use of LMOs resulting from modern biotechnology that may have adverse effects on the conservation and sustainable use of biodiversity, taking into account risks to human health, and focusing on transboundary movements (Article 1)			

5. Conclusion

Currently the relationship between the Biosafety Protocol and the SPS Agreement remains somewhat ambiguous. Although there is considerable interplay between these institutions, it seems unlikely in the short term that the relationship between the Protocol and the SPS Agreement will be formally clarified. However, given the overlapping membership and members' interest in avoiding a conflict between the rules of the biosafety and trade regimes, it is possible to be optimistic and to anticipate that the two regimes' objectives, principles and practices would be interpreted in mutually favourable ways. This practice could bring about a gradual convergence of objectives, principles and rules in the two regimes, leading possibly to a de facto clustering on trade rules for transboundary GMOs. On a more pessimistic note, any convergence will have its price. For example, the inclusion in the Biosafety Protocol of current non-members, such as the United States or Canada, could come at the expense of diluted implementation or compliance measures.

Although there is certainly potential for convergence, the relationship between the two regimes has yet to be defined. The Biosafety Protocol has to date not been formally recognized by WTO members as establishing standards applicable for the SPS Agreement, or as likely to become a standard-setting authority on labelling standards for the TBT Agreement. It is doubtful that this would easily change in the foreseeable future, given that this would strengthen the Protocol vis-à-vis the WTO – something a number of WTO members would be likely to oppose. Additionally, ongoing WTO negotiations on the relationship between trade measures in multilateral environmental agreements and WTO rules, established at the WTO's Doha Ministerial Conference in 2001, will not directly address this relationship.

One open question relates to the role the biosafety regime could play regarding the setting of international safety standards. Although the Codex has the prime responsibility for this, it has not yet developed international health and safety standards relating to transboundary GMOs. There is, at least in theory, a role the Biosafety Protocol could play in this process. The Protocol has mechanisms in place, such as the advance informed agreement and the Biosafety Clearing-House mechanisms, that have a potential role to play in both identifying and collecting information relating to the safety or risk of transboundary LMOs.

Another option is for parties to develop standards in a third forum. The Codex, for example, could provide a forum for developing and adopting safety standards and procedures for GMOs that would be seen as consistent with parties' obligations under the SPS Agreement and the Protocol. The Codex Committee on General Principles is currently developing general principles for risk analysis of GMO imports, and is debating under what conditions precautionary action should be warranted and what criteria should be used to determine whether precautionary action is justified. It is also examining the analytical methods available for detecting GMOs in foods. This should be likely to provide a way forward on establishing safety standards that are acceptable to both importing and exporting countries of LMOs.

The Biosafety Protocol was negotiated because a regulatory gap existed in trade rules governing the transboundary movement of GMOs. Neither the WTO nor any other related body, such as the Codex Alimentarius, had stepped in to fill this gap. The interests of GMO-importing countries – seeking to protect their consumers from possible health risks and their ecosystems from the potentially adverse effects of the growing international trade in GMOs – were no longer adequately represented by the WTO. The Biosafety Protocol has strengthened the argument for trade regulation to protect human, animal and plant health in the biosafety community, but the competencies of the two regimes have not (yet) been sufficiently delineated.

The relationship of these overlapping regimes will quite likely continue to evolve on an ad hoc basis through the strategic practices of their parties, domestic implementation or any dispute settlement proceedings, possibly over time leading to a de facto clustering of the thus far overlapping regimes. Both regimes can be expected to affect one another's performance substantially, making the interplay reciprocal, which would provide incentives to combine the respective principles and rules. It is thus possible that conflict will be averted because the functional linkage between the biosafety and trade regimes will probably cause issue-based political interplay. Member countries in both regimes are likely to be motivated to enhance institutional effectiveness; hence, in the medium to long term, they will tend to seek to reconcile differences in the two regimes' approaches to trade in transboundary GMOs.

Notes

1. See IDGEC's homepage for more details at ⟨http://www2.bren.ucsb.edu/~idgec⟩ (accessed 6 July 2007).
2. See IHDP's homepage for more details at ⟨http://www.ihdp.org⟩ (accessed 6 July 2007).
3. The term "precautionary approach" was adopted at the insistence of the United States, arguing that a "precautionary principle", which is what the Europeans would have preferred, would imply its recognition as a universal principle of international law, which was not the case in the US opinion.
4. The terms GMO and LMO are largely used interchangeably. For an explanation of the roots of this distinction, see Gupta (Chapter 2 in this volume, p. 27) and Gupta (2004: 134–136).

REFERENCES

Alcock, Frank (2002), "Scale Crisis and Sectoral Conflict: The Fisheries Development Dilemma", in Frank Biermann et al. (eds), *Global Environmental Change and the Nation State: Proceedings of the 2001 Berlin Conference on the Human*

Dimensions of Global Environmental Change. Potsdam: Potsdam Institute for Climate Impact Research.

Anderson, T. (2002), "The Cartegena Protocol on Biosafety to the Convention on Biological Diversity: Trade Liberalisation, the WTO, and the Environment", *Asia Pacific Journal of Environmental Law* 7(1): 1–38.

Andrée, Peter (2005), "The Cartagena Protocol on Biosafety and Shifts in the Discourse of Precaution", *Global Environmental Politics* 5(4).

Appellate Body Report (1998), *European Communities – Measures Concerning Meat and Meat Products (Hormone Case)*, WT/DS26/AB/R, WT/DS48/AB/R.

Benedick, Richard (1998), *Ozone Diplomacy: New Directions in Safeguarding the Planet*, revised edn. Cambridge, MA: Harvard University Press.

Brack, Duncan, Robert Falkner and Judith Goll (2003), *The Next Trade War? GM Products, the Cartagena Protocol and the WTO*. Briefing Paper No. 8, London: RIIA, September.

Cartagena Protocol (2000), *Cartagena Protocol on Biosafety to the Convention on Biological Diversity: Text and Annexes*. Montreal: Secretariat of the Convention on Biological Diversity; available at ⟨http://www.cbd.int/doc/legal/cartagena-protocol-en.pdf⟩ (accessed 5 July 2007).

Contreras, Antonio, Louis Lebel and Suparb Pasong (2001), *The Political Economy of Tropical and Boreal Forests*, IDGEC Scoping Report No. 3. Hanover, NH: IDGEC.

Ebbin, Syma (2002), "Enhanced Fit through Institutional Interplay in the Pacific Northwest Salmon Co-management Regime", *Marine Policy* 26(4): 23–29.

Falkner, Robert and Aarti Gupta (2004), *Implementing the Biosafety Protocol: Key Challenges*, Sustainable Development Programme Briefing Paper SDP BP 04/04. London: RIIA, November.

Gupta, Aarti (2000), "Governing Trade in Genetically Modified Organisms: The Cartagena Protocol on Biosafety", *Environment* 42(4): 23–33.

Gupta, Aarti (2004), "When Global Is Local: Negotiating Safe Use of Biotechnology", in Sheila Jasanoff and Marybeth Long Martello (eds), *Earthy Politics: Local and Global in Environmental Governance*. Cambridge, MA: MIT Press, pp. 127–148.

Gupta, Aarti and Robert Falkner (2006), "The Influence of the Cartagena Protocol on Biosafety: Comparing Mexico, China and South Africa", *Global Environmental Politics* 6: 23–55.

Helm, Carsten and Detlef F. Sprinz (2000), "Measuring the Effectiveness of International Environmental Regimes", *Journal of Conflict Resolution* 45(5): 630–652.

Hoel, Alf Hakon, with Elena Andreeva, Russell Reichelt, Virginia Walsh and Oran R. Young (2000), *Performance of Exclusive Economic Zones*, IDGEC Scoping Report No. 2. Hanover, NH: IDGEC.

Isaac, Grant E. and William A. Kerr (2003), "Genetically Modified Organisms and Trade Rules: Identifying Important Challenges for the WTO", *World Economy* 26(1): 29–42.

Kim, Joy A. (2001), "Institutions in Conflict? The Climate Change Flexibility Mechanisms and the Multinational Trading System", *Global Environmental Change* 11(3): 251–255.

Kim, Joy A. (2004), "Regime Interplay: The Case of Biodiversity and Climate Change", *Global Environmental Change* 14(4): 315–324.

King, Leslie A. (1997), "Institutional Interplay: Research Questions. A Report for Institutional Dimensions of Global Change, International Human Dimensions Programme on Global Environmental Change", Draft, University of Vermont, September; available at ⟨http://www2.bren.ucsb.edu/~idgec/publications/idgecscience/InstitutInterplay.pdf⟩ (accessed 6 July 2007).

Lebel, Louis (2005), "Institutional Dynamics and Interplay: Critical Processes for Forest Governance and Sustainability in the Mountain Regions of Northern Thailand", in U. M. Huber, H. K. M. Bugmann and M. A. Reasoner (eds), *Global Change and Mountain Regions: An Overview of Current Knowledge*. Berlin: Springer-Verlag, pp. 531–540.

Miles, Edward L., Arild Underdal, Steiner Andresen, Jorgen Wettestad, Jon Birger Skjaerseth and Elaine M. Carlin (2002), *Environmental Regime Effectiveness: Confronting Theory with Evidence*. Cambridge, MA: MIT Press.

Oberthür, Sebastian (2001), "Linkages between the Montreal and Kyoto Protocols – Enhancing Synergies between Protecting the Ozone Layer and the Global Climate", *International Environmental Agreements* 1(3): 357–377.

Post, Diahanna L. (2005), "Precaution versus Risk in the International Arena: The Case of Codex Alimentarius", paper presented at the Annual Meeting of the International Studies Association, Honolulu, Hawaii, 1–5 March.

Safrin, Sabrina (2002), "Treaties in Collision? The Biosafety Protocol and the World Trade Organization Agreements", *American Journal of International Law* 96(3): 606–628.

Schroeder, Heike (2003), "From Dusk to Dawn: Japan's Climate Change Policy", doctoral dissertation, Free University of Berlin.

Schroeder, Heike, Leslie A. King and Simon Tay (forthcoming), "Contributing to the Science–Policy Interface: Policy Relevance of Findings on the Institutional Dimensions of Global Environmental Change", in Oran R. Young, Leslie A. King and Heike Schroeder (eds), *Institutions and Environmental Change: Principal Findings, Applications, and Research Frontiers*.

Shiva, Vandana (1998), "An Interview with Dr. Vandana Shiva", *Motion Magazine*, 14 August, ⟨http://www.inmotionmagazine.com/shiva.html⟩ (accessed 6 July 2007).

Sewell, Granville, Merrylin Wasson and Yoshiki Yamagata (2000), *The Institutional Dimensions of Carbon Management*, IDGEC Scoping Report No. 1. Hanover, NH: IDGEC.

SPS Agreement (1994), *Agreement on the Application of Sanitary and Phytosanitary Measures. Annex 1A to the Final Act Embodying the Results of the Uruguay Round of Multilateral Trade Negotiations*, Marrakesh, 15 April 1994; available at ⟨http://www.wto.org/English/docs_e/legal_e/15-sps.pdf⟩ (accessed 5 July 2007).

Stokke, Olaf Schram (2004), "Trade Measures and Climate Compliance: Institutional Interplay between WTO and the Marrakesh Accords", *International Environmental Agreements: Politics, Law and Economics* 4: 339–357.

Susskind, Lawrence (1993), *Environmental Diplomacy: Negotiating More Effective Global Agreements*. Oxford: Oxford University Press.

TBT Agreement (1994), *Agreement on Technical Barriers to Trade. Annex to the Final Act Embodying the Results of the Uruguay Round of Multilateral Trade Negotiations*, Marrakesh, 15 April 1994; text available at ⟨http://www.wto.org/english/docs_e/legal_e/17-tbt.pdf⟩ (accessed 5 July 2007).

Underdal, Arild (2002), "Patterns of Regime Effectiveness", in Edward L. Miles et al. (eds), *Environmental Regime Effectiveness: Confronting Theory and Evidence*. Cambridge, MA: MIT Press, pp. 433–465.

Underdal, Arild and Oran R. Young, eds (2004), *Regime Consequences: Methodological Challenges and Research Strategies*. Dordrecht: Kluwer Academic Publishers.

Vogel, David (1997), "Trading up and Governing across: Transnational Governance and Environmental Protection", *Journal of European Public Policy* 4(4): 556–571.

Winham, Gilbert R. (2003), "International Regime Conflict in Trade and Environment: The Biosafety Protocol and the WTO", *World Trade Review* 2: 131–155.

Young, Oran R. (1994), "The Problem of Scale in Human/Environment Relations", *Journal of Theoretical Politics* 6: 429–447.

Young, Oran R. (1996), "Institutional Linkages in International Society", *Global Governance* 2: 1–23.

Young, Oran R. (1999a), *Governance in World Affairs*. Ithaca, NY: Cornell University Press.

Young, Oran R., ed. (1999b), *The Effectiveness of International Environmental Regimes*. Cambridge, MA: MIT Press.

Young, Oran R. (2001), "Evaluating the Effectiveness of International Environmental Regimes: Where Are We Now?", *Global Environmental Politics* 1: 99–121.

Young, Oran R. (2002a), *The Institutional Dimensions of Environmental Change: Fit, Interplay, and Scale*. Cambridge, MA: MIT Press.

Young, Oran R. (2002b), "Matching Institutions and Ecosystems: The Problem of Fit", Institut du Développement Durable en Relations Internationales (IDDRI), Paris.

Young, Oran R. (2002c), "Institutional Interplay: The Environmental Consequences of Cross-Scale Interactions", in Elinor Ostrom et al. (eds), *The Drama of the Commons: Institutions for Managing the Commons*. Washington, DC: National Academy Press.

Young, Oran R. (2003a), "Environmental Governance: The Role of Institutions in Causing and Confronting Environmental Problems", *International Environmental Agreements* 3: 377–393.

Young, Oran R. (2003b), "Regime Effectiveness: A Commentary on the Oslo-Potsdam Solution", *Global Environmental Politics* 3: 97–104.

Young, Oran R., with contributions from Arun Agrawal, Leslie A. King, Peter H. Sand, Arild Underdal and Merrilyn Wasson (1999), *Science Plan: Institutional Dimensions of Global Environmental Change*, IHDP Report No. 9. Bonn: International Human Dimensions Programme on Global Environmental Change. Revised edn prepared by Heike Schroeder, IHDP Report No. 16, 2005; available at ⟨http://www2.bren.ucsb.edu/~idgec/publications/IHDP-IDGECreport16.pdf⟩ (accessed 6 July 2007).

4

Overlapping regimes: The SPS Agreement and the Cartagena Biosafety Protocol

Are K. Sydnes

1. Introduction

This chapter addresses the institutional causes of the conflict between the environment and trade in the case of biosafety; that is, the institutional interplay between the WTO agreements – the General Agreement on Tariffs and Trade (GATT), the Agreement on the Application of Sanitary and Phytosanitary Measures (SPS Agreement) and the Agreement on Technical Barriers to Trade (TBT Agreement) – and the Cartagena Protocol on Biosafety of the Convention on Biological Diversity (CBD), which regulates the transboundary movement of genetically modified organisms (GMOs). I discuss the nature of the institutional interplay between the two agreements as they co-evolve, and the extent to which this interplay may be synergetic or may impede the effectiveness of the individual regimes and the global regulation of the transboundary movement of GMOs.

My analytical aim is to address the issue of overlapping regimes.[1] To what extent do the norms, rules and procedures of the WTO agreements and the Cartagena Protocol overlap, and, if they do, are they compatible? What rules are to prevail in a case where a dispute arises regarding the interpretation and compatibility of the WTO agreements and the Cartagena Protocol? How do states seek to manage the potential overlaps between the two regimes? Do overlaps between the two regimes impede or promote the effectiveness of the global regulation of biosafety? In order to understand the institutional causes and effects of interplay, their causal

Institutional interplay: Biosafety and trade, Young, Chambers, Kim and ten Have (eds), United Nations University Press, 2008, ISBN 978-92-808-1148-3

mechanisms need to be addressed: to what extent and how do the contents, operations and consequences of one regime affect those of another? I apply an analytical approach to identify the causal pathways of interplay, distinguishing between political, normative and operational interplay (Stokke 2001; Rosendal 2001a, 2001b). My aim is to analyse the consequences of interplay for the effectiveness of the Cartagena Protocol and the WTO regime and the global regulation of biosafety.

For effectiveness, an analytical distinction can be made between the outputs, outcomes and impacts of international institutions (Underdal 2004). Outputs relate to the rules and norms established by an institution – for example, the Protocol and the decisions of its Meeting of Parties, the WTO agreements and the rulings of WTO panels and Appellate Bodies. Outcomes denote behavioural effects of the institution on relevant actors. The question here is whether parties to the institutions alter their behaviour as a consequence of the institutional arrangement. Finally, impact relates to the environmental effects of an institution, or, more broadly, its effects on the issue it was established to alleviate (e.g. conservation, trade, management or security).

In the case of the global biosafety regime and the interplay of the Protocol and the WTO agreements, it is premature to arrive at any general conclusions with regard to their impact on the regulation of the transboundary movement of GMOs. However, it is possible to analyse institutional aspects and how these may affect the behavioural patterns of states (outcomes) and the outputs of the regimes as these are moulded by their interactions. An underlying question in this regard is: who stands to gain or lose from the different results of institutional interplay?

2. Analysing institutional overlap

International regimes are social institutions that define practices, assign roles and guide the interactions of the occupants of these roles within an issue area (Young 1996: 3). Regimes commonly comprise a substantive component of rights and rules and an operative component establishing procedures facilitating their establishment and implementation (Stokke 2001). The study of international regimes, perceived of as functionally and spatially circumscribed within given issue areas, has been a prevalent feature of international relations theory over recent decades (Stokke 1997). As a research programme, regime theory has provided valuable insights and new knowledge on international cooperation. However, the focus on specific issue areas has limited the analytical scope of regime theory in terms of analysing the interactions of individual international institutions and regimes. Institutional interplay refers to a situation where the contents, operations or consequences of one institution affect those of another (Stokke 2001). This volume on institutional interplay is an effort to enhance the understanding of this phenomenon.

2.1. Institutional overlaps

Young (1996, 2002) identifies four forms of institutional linkage: embedded, nested, overlapping and clustered. This typology refers to the structure of the relation between two or more institutions. Institutional linkages are, thus, empirical and relational, depending on the individual case. Institutional interplay, on the other hand, is used to denote the process of such interactions and their outcomes at the actor or institutional level.[2] The international regulation of biosafety, the case at hand, falls to varying extents within the scope of several international agreements and institutions.[3] Such institutional overlaps are an emerging, though under-studied, field of research. Initial efforts have been made by Young (1996), Young et al. (1999) and Rosendal (2001a, 2001b) to outline an analytical approach to such phenomena. Young has defined overlapping regimes as cases where "individual regimes that were formed for different purposes and largely without reference to one another intersect on a de facto basis, producing substantial impacts on each other in the process" (1996: 6). Young's definition underscores the functional intersections of individual regimes that were largely established without reference to the other(s). However, the qualification that regimes are established "largely without reference to one another" seems more restrictive than need be for analytical purposes. In the case of the Cartagena Protocol on Biosafety, the negotiators were aware from the start of the negotiations that there was ample room for overlap with the SPS Agreement and other WTO agreements. Indeed, disagreements during the negotiation of the Protocol were clearly driven by the participants' perception of overlaps between the Protocol and the prevailing regulation of trade in GMOs under the WTO. Rosendal proposes a more general definition of institutional overlaps that seems more appropriate for analysing the case of biosafety. "Regimes are overlapping when their policy goals and regulations prescribed for problem solving intersect within the same issue area" (Rosendal 2001b: 458). Rosendal's definition does not stress the relative autonomy of the formation processes of the individual regimes, as is the case with Young's definition.

Institutional overlap is often considered as an externality,[4] that is, an unintended consequence of institutional developments within separately defined issue areas. However, overlaps can also be intentional, a result of strategic moves (Rosendal 2001b: 458).[5] The definition and delineation of institutional scope and the determination of whether institutional overlap exists may be politically contentious (Rosendal 2001b: 458). Indeed, commentators disagree on whether there is a substantive overlap between the Cartagena Protocol and the relevant WTO agreements (Kerr and Phillips 2000; Rivera-Torres 2003). If there is, to what extent will such an overlap impede or enhance the efficiency of the global regime for the transboundary movement of GMOs?

Institutional overlaps may be dealt with in different manners by the institutions and parties involved. The regimes or the agreements establishing them may themselves set rules of precedence between international regimes.[6] The overlap may also be resolved in legal terms based on rules of precedence according to international law. However, in many cases the overlap between regimes is not resolved by legal ruling or codification, but is subject to political interpretation and negotiations. In some cases, parties may seek to coordinate the overlapping regimes to avoid the – often costly – duplication of functions. In such cases, coordination is motivated by the prospect of generating synergies. In other cases, institutional overlaps may lead to "turf wars" where parties consider one or the other regime to be most appropriate to deal with a specific issue at hand. In such instances, one may expect "forum shopping" and disputes to be prevalent. In general, the management of interplay requires that there is awareness among the participants of the institutional overlap and a deliberate intention to manage the effects of the interplay between them (Stokke 2001: 11).

In which cases do institutional overlaps become malignant or benign? To address this issue it is useful to unbundle the substantive nature of institutional overlaps. First, a distinction can be made between institutional overlaps that pertain to core and secondary aspects of regimes (Rosendal 2001a: 100). *Core aspects* constitute the defining features of a regime – for example, principles relating to the sharing of benefits and the importance ascribed to trade or environmental protection. As such, core aspects commonly have distributive consequences among the parties to the agreement. *Secondary aspects* are more relative in nature – for example, the seriousness of a particular problem to be addressed or the importance of causal factors.[7] Rosendal (2001a) also distinguishes between *regulatory rules* and *programmatic rules*. The former establish the rights, duties and obligations that parties are expected to act upon. The latter are efforts under the regime to, for example, increase the knowledge basis within an issue area, the development of techniques, and so forth.

On this basis it is reasonable to argue that discord over core aspects and regulatory rules has a greater potential for conflict in the case of overlapping regimes. Secondary aspects and programmatic rules, on the other hand, may prove more conducive to coordination and synergetic interplay. By applying this typology of overlaps one may be able to distinguish to a greater extent between the different aspects of regimes that cause synergy or interference in institutional interplay.

2.2. Causal mechanisms

The nature of overlaps has substantial consequences for the process of institutional interplay, its outcomes, the effectiveness of the individual regimes and, more broadly, the regulation of the issue area in question. In this section I address the causal pathways of institutional interplay. Causal pathways are analytical categories for analysing the forces that structure the process of institutional interplay. Stokke (2000) identifies three such causal pathways by which one regime may influence another: normative, political and operational interplay (Rosendal 2001b: 459).[8]

Normative interplay denotes the diffusion of regime features, such as principles or regulations, and their adoption by other regimes. That is, decision-making procedures, allocation principles, environmental standards or other regime components that have proved efficient or gained substantial legitimacy within the framework of one regime may be adopted by other regimes (Stokke 2000). This has, for example, been the case when regional fisheries regimes have adopted the organizational designs and practices of other similar regimes (Stokke 2000; Sydnes 2002). Such interplay has proved to be enhanced when there is closeness in time, space, participation or functional orientation between the regimes involved (Stokke 2000: 225). However, it is also recognized that normative interplay becomes more difficult in cases where it may have distributional effects among the members of a recipient regime (Stokke 2000). As such, normative interplay is politically sensitive and is most prevalent where there is broad consensus and/or low-cost decisions are involved.

The second causal pathway identified by Stokke (2001) is political interplay. This implies that the interests and capabilities established by one regime spill over into a recipient regime. Take the case of the United Nations Convention on the Law of the Sea of 1982 (UNCLOS) and its consequences for the functions of established regional (multilateral) fishery regimes. By giving coastal states sovereign rights over an ocean area of 200 nautical miles from their coasts, UNCLOS in effect circumscribed the geographical scope and management authority of established regional fishery regimes in the Atlantic Ocean and other regions (Sydnes 2001; 2005: 121–123). Where overlapping regimes have incompatible provisions on the distribution of rights and duties, one may expect political interplay to be contested. In some cases such disputes may be settled legally by rules of precedence; in other cases there is a need for political resolution. It may also prove more difficult to make political interplay relevant across issue areas (Stokke 2000: 227) than within an established issue area. This is the case with biosafety and more generally with trade and the environment. Here, that rights and duties established under one regime should hold in the other is hotly debated.

Whereas the first two pathways of interplay revolve around the substantive features of international regimes, operational interplay addresses the procedural aspects of regime interplay. Here action is taken to coordinate, or make compatible, the norms and procedures of the regimes involved (Stokke 2000: 230). The motivation for operational interplay may be to avoid wasteful duplication or to pool (often scarce) technical, human or financial resources. Operational interplay may also be a way to ensure the normative coherence of overlapping regimes, for example by establishing common institutional mechanisms for knowledge production. As noted by Rosendal (2001b: 459), this pathway seems restricted to cases where there is an actual functional overlap between regimes, and would require institutionalized mechanisms for coordination among the regimes. Operational interplay can, as such, be regarded as institutional adaptation to the other forms of interplay noted above (Stokke 2000: 230).

When analysing cases of institutional overlap (and interplay more broadly), it is important to distinguish not only the impact of individual causal mechanisms but also whether these operate jointly and have an internal dynamic. One could, for example, imagine how allocation issues (political interplay) could become less contentious through the diffusion and sharing of technical regulations (operational interplay) and environmental standards (normative interplay). However, one could equally imagine the impeding dynamics of normative, political and operational interplay between overlapping regimes in cases where core values and norms have distributional consequences.

By applying Rosendal's (2001a) typology of regime overlaps and Stokke's (2001) causal mechanisms one may formulate the following propositions:

- The core aspects and regulatory rules of regimes are more politically sensitive than other types of overlap. They constitute the defining features of regimes and the rights and duties of members, and their degree of compatibility will be pivotal in determining whether the interplay between them will be driven by synergy or by interference. In such cases, political interplay will come into play, either as synergetic or as impeding to the effectiveness of the regimes.
- Normative interplay is most benign when the core aspects and regulatory rules of regimes are compatible. Such interplay is further enhanced where there is closeness in time, overlap of membership and a similarity in functions between the regimes (Stokke 2001). In general,

the secondary aspects of regimes are more easily diffused than are other regime features, because they have fewer distributive consequences.
• Programmatic regulations are more benign to operational interplay between regimes than are other substantive or operational aspects of institutional overlap. However, since operational interplay can be seen as an institutional adaptation of the other causal pathways, it is to an extent conditioned by these (Rosendal 2001a).

3. The WTO agreements and the Cartagena Protocol on Biosafety

3.1. The WTO agreements

The WTO has evolved over a period of more than 50 years, and has its roots in the 1947 General Agreement on Tariffs and Trade (GATT). The aim of GATT was the dismantling of barriers to international trade in goods. This has been achieved through expansion in terms of issue areas, multilateral decision-making and the rulings of an efficient dispute settlement mechanism with the authority to impose trade sanctions (Cottier 2002: 468). The substantive content of the WTO regime lies in the portfolio of agreements negotiated under its auspices. As described in the previous chapters, of particular relevance to biosafety governance and the interplay with the CBD's Cartagena Protocol on Biosafety are the GATT agreements of 1947/1994, the Agreement on Technical Barriers to Trade of 1994 (TBT Agreement), but primarily the Agreement on the Application of Sanitary and Phytosanitary Measures of 1994 (SPS Agreement).

GATT requires that members give equal treatment to exports from all member countries (Article I) and that there is no discrimination between locally produced and imported products (Article III). GATT contains a general exceptions clause that provides the basis for environmental measures in international trade, on condition that such measures are not set arbitrarily and do not constitute unjustifiable discrimination or a disguised restriction (GATT, Article XX: Chapeau).

The SPS Agreement is a trade agreement applying to the "unjustified" use of national sanitary (human and animal health) and phytosanitary (plant health) measures in international trade. The approach of the SPS Agreement is to establish rules and procedures on which national sanitary and phytosanitary measures are to be based. Moreover, it seeks to harmonize such measures through the use of international standards established by international organizations, the Codex Alimentarius Commission in the case of GMOs (Boutrif 2003). The balancing act of the

SPS Agreement is to allow for those regulatory measures necessary for the protection of human, animal and plant health, while preventing measures that represent arbitrary or unjustifiable restrictions on international trade. This is done through rules and procedures on scientific risk assessment and management (Article 5).

Regulatory measures applying to GMOs that do not fall under the SPS Agreement – for example measures that are introduced to protect biodiversity or other kinds of health and safety issues – may in some cases be subject to the TBT Agreement (Howse and Meltzer 2002: 491).[9] The aim of the TBT Agreement is to prevent members applying technical standards and regulations – for example on packaging and labelling – as barriers to international trade (Article 2.2).

3.2. The Cartagena Protocol on Biosafety

The negotiation of the Cartagena Protocol on Biosafety was initiated on the basis of Article 19.3 of the Convention on Biological Diversity (CBD), which provided for the development of a legally binding international instrument on biosafety. The Protocol was one of the first multilateral environmental agreements (MEAs) to implement the precautionary approach, as defined by Principle 15 of the Rio Declaration on Environment and Development (Article 1), although the term itself does not appear in the provisions of the agreement. The main tool to determine acceptable levels of risk and uncertainty is the application of risk assessment and risk management (Articles 15–16, Annex III). Importantly, in cases where there is insufficient scientific information and knowledge, the importing party may apply precaution (Articles 10.6 and 11.8).

The advance informed agreement (AIA) procedure is frequently referred to as the centrepiece of the Protocol (Gupta 2001). It implies that there is an obligation on exporting countries to solicit an AIA prior to the transfer of a GMO intended for deliberate release into the environment. For other categories of GMOs – those for contained use (laboratories), agricultural commodities (i.e. food, feed or for processing) or processed products derived from GMOs – there is no obligation for prior consent or information-sharing, only that such information (to varying degrees) accompanies the transfer. Chapter 2 gives more detail on the workings of the system; suffice it to say here that one of the innovative mechanisms established by the Protocol is the Biosafety Clearing-House for information-sharing. An exporter country is obliged to notify the Biosafety Clearing-House 15 days before the approval of the introduction of a GMO into its own agricultural production. This allows potential countries of import to consider whether they want to conduct risk assessments or prohibit imports of the GMO on the grounds of precaution.

In contrast to the WTO, which is a multilateral organization with supranational authority, the Protocol is more of a traditional MEA, enabling member countries to establish domestic regimes for the regulation of biosafety (Cottier 2002: 468).

3.3. Overlap and the rules of precedence

Delimiting the scope of international institutions pertaining to the same issue area or determining whether there is an overlap between them may be both politically contentious and analytically challenging. As already noted, commentators have taken different views on whether there actually is an overlap between the WTO agreements and the Cartagena Protocol, and whether or not the provisions of the agreements in question are compatible. An initial question to be addressed in the analysis is therefore how the rules of precedence apply in the case of biosafety. This was one of the main issues of contention during the negotiation of the Protocol. Which agreement was to take precedence in a case concerning the transboundary movement of GMOs: the rules of the WTO agreements or those of the Cartagena Protocol? As discussed by Aarti Gupta in Chapter 2, this question led to the following preamble for the Protocol:

> *Recognizing* that trade and environment agreements should be mutually supportive with a view to achieving sustainable development,
>
> *Emphasizing* that this Protocol shall not be interpreted as implying a change in the rights and obligations of a Party under any existing international agreements,
>
> *Understanding* that the above recital is not intended to subordinate this Protocol to other international agreements.

The implications of this formulation in the preamble have been debated widely. Some claim they "muddy the waters" on biosafety governance (Kerr and Phillips 2000; Charnovitz 1999/2000). Others argue that it reflects the negotiators' awareness of the interdependence of the trade and environmental aspects of GMOs. The formulation on the mutual supportiveness of trade and environmental agreements clearly reflects the differing intentions of the negotiating parties.[10,11] Gupta reaches the conclusion that the outcome would be conditioned by political factors: "The unsurprising outcome is that, far from resolving the trade–environment conflict, these seemingly contradictory statements will be interpreted to suit different needs" (Gupta 2000: 31). If anything, the formulation of the preamble of the Cartagena Protocol has levelled the playing field between it and the relevant WTO agreements. There is little likelihood that the nature of the overlap will be resolved through a legal ruling or according to rules of precedence in international law. Hence, the nature of the overlap and interplay between the Cartagena Protocol and the WTO agreements will have to be resolved by the state parties.

4. Institutional overlap in practice: Conditions for introducing regulatory measures

To discuss the substantive nature of the overlap between the Cartagena Protocol and the WTO agreements, this section addresses a core issue at stake in the debate: the conditions prescribed for introducing trade regulations on GMOs under the two regimes. This debate is taking place against the background of a high level of scientific uncertainty regarding the potential effects of GMOs on the natural environment and on human health (Busch et al. 2004). Nonetheless, both the SPS Agreement and the Cartagena Protocol rely on the methods of scientific risk assessments to deal with uncertainties related to the transboundary movement of GMOs. A possible dispute between the WTO agreements and the Cartagena Protocol relates to the environmental principles underlying decisions made on the basis of risk assessments. Both the SPS Agreement and the Cartagena Protocol require that regulatory decisions regarding the introduction of GMOs are to be based on science-based risk assessments. However, there is no international consensus on the methodological and scientific standards for such assessments.[12] As a result, the scientific and environmental standards established by the Cartagena Protocol and the SPS Agreement, respectively, have become a matter of discussion. I present a brief outline of the rules and procedures for imposing regulatory measures on GMOs, followed by a discussion of the extent to which these are compatible.

The SPS Agreement states that sanitary and phytosanitary measures should be applied "only to the extent necessary to protect human, animal or plant life or health". Such measures are to have a scientific justification and are to be based on sufficient scientific evidence (SPS Agreement: Article 2.2). The means of reaching such a conclusion is risk assessment (Article 5.1). According to the SPS Agreement, risk assessments shall include the "evaluation of the likelihood of entry, establishment or spread of a pest or disease within the territory of an importing Member ... and of the associated potential biological and economic consequences; or the evaluation of the potential for adverse effects on human or animal health" (Annex A.4). Before applying a regulatory measure, an importing country must decide upon its desired level of protection. In doing so it must seek to minimize the negative effects on trade (Article 5.4) and, further, ensure that the level of protection is consistent with avoiding arbitrary distinctions (Article 5.5). The regulatory measure must be designed so as to ensure that it is not more trade restrictive than required. Measures established by other countries, though they may differ from the established national standard, are to be deemed equivalent if they are proved to have the same effects (Article 4.1). In addition to these requirements, the WTO Appellate Body in the Japan Agricultural Products Case added the requirement that there be a rational or objective relationship between the measures imposed and the scientific evidence available (Rivera-Torres 2003: 299). Consequently, the introduction of regulatory measures under the SPS Agreement is subject to a number of conditions. The SPS Agreement nonetheless does allow for a level of precaution to be taken by importing countries in cases of scientific uncertainty.

> In cases where relevant scientific evidence is insufficient, a Member may provisionally adopt sanitary or phytosanitary measures on the basis of available pertinent information, including that from the relevant international organizations as well as from sanitary or phytosanitary measures applied by other Members. In such circumstances, Members shall seek to obtain the additional information necessary for a more objective assessment of risk and review the sanitary or phytosanitary measure accordingly within a reasonable period of time. (SPS Agreement: Article 5.7)

The SPS Agreement thereby allows for measures to be adopted in a precautionary approach based on pertinent information and on a provisional basis. However, there is an obligation on the implementing party to obtain the necessary information within a reasonable time frame to conduct a more objective risk assessment. This may be deemed a conditioned application of the precautionary approach where the burden of proof is carried by the country of import. This, in essence, reflects the regulatory approach of the SPS Agreement: to allow for regulatory measures in cases where these are needed but to ensure that such measures are not introduced in an arbitrary manner representing discrimination and disguised restrictions on international trade (SPS Agreement: Preamble).

The SPS Agreement also allows for a degree of national flexibility in the regulatory measures adopted. However, there are conditions for the establishment of such measures by importer countries. Members of the WTO are to base their sanitary and phytosanitary measures on international standards, guidelines or recommendations (Article 3.1). In the case of biosafety, the relevant international organization establishing such standards is the Codex Alimentarius Commission (Article 3.4). National SPS measures that are stricter than those established by international standards need to be scientifically justified (Article 3.3). These conditions imply that the scope for national variations is limited in terms of their regulatory effects.

In cases where the regulation of GMOs does not fall under the scope of the SPS Agreement, such as for the purpose of biodiversity or other kinds of health and safety purposes, it will in most cases come under the scrutiny of the TBT Agreement (Howse and Meltzer 2002: 491). A fundamental aim of the TBT Agreement is to prevent technical standards and regulations from being prepared, adopted or applied in an arbitrary or discriminatory manner, creating unnecessary obstacles to international trade.

Scientific risk assessment is also at the heart of the Cartagena Protocol. Risk assessment, as outlined by Article 15 and Annex III, is to provide the scientific basis for the intentional introduction of GMOs into the environment by an importing country (Article 10.1). In addition, the procedures for notifying the Biosafety Clearing-House about the import of GMOs for feed, food or processing (Article 20) require that the information submitted includes a risk assessment consistent with Annex III (Andren and Parish 2002: 329). Risk assessments are to identify and evaluate the possible adverse effects of a GMO on biodiversity, also taking into account human health. They are to be conducted in a "scientifically sound" and transparent manner, according to recognized risk assessment techniques (Article 15.1; Annex III). It is recognized that risk assessments need to be carried out on a case-by-case basis (Annex III.6). Importantly, the costs of conducting a risk assessment may be borne by the notifier (exporter) if the country of import so requires (Article 15.3).

Rivera-Torres argues that "it would be difficult to argue that a risk assessment carried out pursuant to the Biosafety protocol would not satisfy the requirements of an 'adequate' risk assessment by the SPS" (2003: 314). However, underlying the rules and procedures on risk assessment in Article 15 and Annex III of the Cartagena Protocol is the precautionary approach.

> Lack of scientific certainty due to insufficient relevant scientific information and knowledge regarding the extent of the potential adverse effects of a living modified organism on the conservation and sustainable use of biological diversity in the Party of import, taking also into account risks to human health, shall not prevent that Party from taking a decision, as appropriate, with regard to the import ... in order to avoid or minimize such potential adverse effects. (Cartagena Biosafety Protocol: Articles 10.6/11.8)

Although the Cartagena Protocol establishes rules and procedures for the conduct of scientific risk assessments as a basis for introducing regulatory measures that are largely compatible with the SPS Agreement, the burden of evidence is reversed. The approach to trade barriers in the SPS Agreement is "why?", whereas the precautionary principle of the Cartagena Protocol, in contrast, asks "why not?" (Kerr and Phillips 2000: 72). The conditions for imposing regulatory trade measures by importing countries and the conditions for applying the precautionary approach as the basis for such decisions are more restrictively formulated in the SPS Agreement than is the case in the Cartagena Protocol. Because uncertainties still prevail in the scientific methods of risk assessment, and political disagreement remains regarding the appropriate levels of risk and uncertainty (Busch et al. 2004), this difference in regulatory approach may hamper the potential for synergetic interplay between the SPS Agreement and the Cartagena Protocol. GMO-exporting nations regard the environmentally sceptical SPS Agreement as the most appropriate domain for the regulation of GMOs on the basis of, primarily, the international standards of the Codex Alimentarius. Meanwhile, predominantly GMO-importing countries may lean towards the Cartagena Protocol, which gives them a larger degree of discretion in establishing domestic regulation of GMOs.

5. Causal pathways and the effectiveness of overlapping regimes

Institutional interplay may have either synergetic or impeding consequences for regime effectiveness. In general it may be assumed that, in the majority of cases, there is substantial potential for synergetic interplay between overlapping regimes (Rosendal 2001a; Stokke 2001). However, whether this potential is realized depends on the substantive nature of the overlap between the regimes (Rosendal 2001b) and how it is managed by the states involved (Stokke 2001). As proposed earlier, the occurrence and consequences of political, normative and procedural interplay are dependent on whether overlaps concern core or secondary aspects of the regimes and whether these are regulatory or programmatic in nature.

Institutional interplay needs to be facilitated by actors, commonly state parties to both the source and recipient regime (Stokke 2001). For example, through a state with overlapping membership, the generation of new knowledge or ideas may diffuse from one regime and influence the decisions made by another regime. Such normative interplay may have substantial synergetic effects and lead to the consistent regulation of an issue area. However, there is little doubt that the outputs of a source institution may have negative effects on a recipient institution. Members of a source institution may establish regulatory rules that are not compatible with those of the recipient institution. In the case of overlapping institutions, and in particular in the case of those co-evolving within the same issue area, such circumstances may result in "turf wars" (Rosendal 2001a and 2001b). The establishment of common scientific standards, negotiating compromises on the establishment of commitments – rights and duties – or establishing coordinating mechanisms may then give rise to difficult disagreements.

Some commentators have stated that, on the face of it, the Cartagena Protocol and the WTO agreements seem compatible. On the basis of a thorough analysis of the scopes of the Cartagena Protocol and of the GATT, TBT and SPS Agreements, Rivera-Torres concludes that there is in fact very limited overlap between the institutions, and further that, where there is overlap, the provisions of the agreements are not in conflict (Rivera-Torres 2003; see also Safrin 2002). Oberthür and Gehring draw a similar conclusion in claiming that remaining tensions stem from the scope for interpretation of both rule systems rather than from obvious incompatibilities of their rules (see Chapter 5 in this volume). Kerr and Phillips, on the other hand, conclude that the Cartagena Protocol is in direct conflict with WTO principles and practices on four counts (2000: 69–74): by allowing for trade barriers on the basis of production and processing methods; by including the precautionary principle as a decision criterion for imposing import bans; by allowing socio-economic factors to be taken into consideration in the decision to approve imports (Article 26); and by labelling requirements for GMOs intended for food, feed or processing.

These disparate accounts of the question of overlap between the two regimes clearly reflect the political uncertainties regarding the global regulation of GMOs. Moreover, because of the unclear rules of precedence, this opens the potential for political conflict. Should the issue of GMOs be subject primarily to the WTO trade regime through the SPS Agreement or to the environmental regime of the CBD as codified by the Cartagena Protocol? This is an issue of significance to states and to economic and political groups with vested interests – of a political or economic nature – in the issue. The disagreements between the "champions" of the agreements may be that the regulation of GMOs is an evolving area of international regulation and institutional development and is subject to considerable political uncertainty. The unresolved question of institutional overlap is a consequence of states seeking to place the process of regulatory development either within the context of the WTO or under the institutional mechanisms established by the CBD and the Cartagena

Protocol. In other words, institutional "forum shopping" is motivated by states seeking to manage the interactions of the regulatory frameworks, acknowledging that the different approaches of the institutions will have an effect on their rights and duties to regulate GMOs through the design of national versus multilateral institutional frameworks.

According to this view, the dominant causal mechanism is that of political interplay, that is, managing the extent to which the rules and norms of one institution affect the utilities of the same state under the regulatory scope of another institution (Stokke 2001). For example, if the WTO, under the SPS and/or TBT agreements, establishes international standards regulating the transfer of GMOs, this would limit the scope of opportunities of states that are both members of the WTO and parties to the Cartagena Protocol to establish regulatory measures according to the environmental standards established by the Protocol. The transatlantic beef hormone dispute between the United States and the European Union is an important driving force in that respect, as was pointed out in Chapter 2. The European Union, although an active party to the SPS Agreement, has an interest in furthering the Cartagena Protocol as the institutional mechanism for regulating biosafety (see Rosendal 2005 for detail). The original members of the Miami Group, on the other hand, would favour the WTO for this task. Developing countries, despite having increasingly diverse national interests as importers and exporters of GMOs, generally have an incentive to support the Cartagena institutions because they provide greater importer-country discretion in formulating national policies and regulations, and exporter countries or the roster of experts may be relied on to cover the costs of risk assessments.

As long as there is no consensus on the functional delineation or the nature of the institutional overlap between the regimes and which agreements are to have precedence in cases of incompatible provisions, there is reason to believe that uncertainties related to the outcomes of such processes of interplay will fuel rather than overcome the impeding nature of the institutional interplay. This is further spurred on by the ongoing transatlantic trade dispute between the United States and the European Union.

In terms of normative interplay, Rivera-Torres (2003) claims there is nothing that formally prohibits the Cartagena institutions from becoming environmental standard-setters for the TBT Agreement.[13] That is, knowledge and ideas could be diffused through normative interplay from one regime to the other, through more or less institutionalized mechanisms (operational interplay). Moreover, the provisions of the SPS Agreement (read in the context of the rulings of the Appellate Body) and of the Cartagena Protocol seem generally compatible as regards the important issues of risk assessment, risk management and the precautionary approach (Rivera-Torres 2003). However, the SPS Agreement dictates that GMO-importing states introducing regulations have a duty to provide scientific analysis within a reasonable time. The question is: what actions are dictated, within what time frame, and what are the conditions for applying a precautionary approach? It seems clear that, under the SPS Agreement, the country of (potential) import carries the burden of evidence and has to qualify its decision to prohibit a GMO.[14] As such it is a conditioned application of the precautionary approach, in contrast to the Cartagena Protocol. These factors also imply that the normative interplay between the two regimes is hampered by member countries perceiving that the diffusion of principles or regulations would have distributional consequences, in terms of affecting established rights and duties. The differences in the precautionary approach between the two agreements may at first glance not seem too great to overcome. However, the precautionary approach is a core aspect of the Cartagena Protocol and is to be applied at the discretion of countries of import (unilaterally). The precautionary approach as applied by the SPS Agreement is a secondary aspect of the regime, to be applied according to multilateral standards. Hence, the precautionary role is given different "weight" within the framework of the two agreements. Moreover, the differences in approach have distributive consequences for countries of import and export. As the example above demonstrates, normative interplay becomes politically sensitive in cases where the diffusion of knowledge, ideas or institutional designs is perceived to have distributive consequences.

There are different scenarios on the outcome of the institutional overlaps between the two regimes and how this will affect the effectiveness of the individual regimes and the regulation of biosafety in general. The members of the WTO and parties to the Cartagena Protocol may reconcile the overlaps of the two regimes, which in turn would provide for substantial synergies in terms of operational interplay through the coordination and/or harmonization of institutional norms and rules. One alternative is to amend GATT Article XX and allow trade measures under specific environmental agreements to be deemed legitimate restrictions on international trade. Such a provision has been written into the North American Free Trade Agreement (Article 104) concerning its interplay with the Convention on International Trade in Endangered Species of Wild Fauna and Flora (CITES), the Montreal Protocol and the Basel Convention (Brack and Gray 2003: 35–36). This would, however, imply that the Cartagena Protocol would take precedence over the relevant WTO agreements in the regulation of biosafety. The Protocol provisions could be applied as evaluation standards or guidelines for decisions and cases under WTO dispute settlement procedures, in particular in cases where the SPS Agreement, the Codex Alimentarius and the TBT Agreement provide limited guidance. In such cases, the Cartagena Protocol, and potentially the Biosafety Clearing-House, would be ascribed the role of establishing international standards. It is, however, unclear how this would comply with the conditions of Article 5.7 of the SPS Agreement.

Alternatively, member states of the institutions in question may resolve functionally to differentiate tasks, that is, to unbundle the overlaps, between the two regimes. Some commentators have proposed that an opportunity for synergetic interplay lies in the WTO-maintained power to assess whether a trade measure is arbitrary, discriminatory or protectionist, whereas the Cartagena Protocol regime retains jurisdiction on the legitimacy, proportionality and necessity of such a measure (Brack and Gray 2003: 35). This would imply a delineation of authority regarding the different stages and/or aspects on a case-by-case basis.

At the present stage of evolution of a global regulatory regime for the transboundary movement of GMOs it may seem that the institutional mechanisms established by the Cartagena Protocol (the Biosafety Clearing-House and the Conference of Parties) and the WTO (the dispute settlement mechanism) constitute effective barriers to the creation of synergies. No one will gain if the disagreements regarding the global regime on biosafety prevail. The current conditions mean that decisions under the Cartagena Protocol – for example a ban on the import of GMOs for food, feed or processing based on the precautionary approach – could be contested under the SPS Agreement and brought before the WTO dispute settlement mechanism. On the other hand, an international standard established by the Codex Alimentarius could hypothetically be disregarded by member states of the Protocol that opt for a higher level of protection.

6. Concluding remarks

The roots of disagreements over the regulation of biosafety are to be found in the history and institutional contexts of the SPS Agreement and the Cartagena Protocol on Biosafety. The Protocol is a part of the CBD and, as such, environmental conservation and the protection of biodiversity are core values of the regime. The SPS Agreement was negotiated under the auspices of the WTO as a trade agreement. The Protocol seeks to avoid detrimental effects on biodiversity (including human health), the SPS Agreement detrimental effects on international trade. These differences are fuelled by the regimes focusing primarily either on exporting countries (the SPS Agreement) or on importing countries (the Cartagena Protocol). Exporting countries see their national interests primarily preserved through the multilateral approach of the SPS Agreement, relying on the Codex Alimentarius Commission to establish international standards for sanitary and phytosanitary measures. The Cartagena Protocol is more of a traditional environmental agreement in appointing the member (importing) countries themselves to decide upon the appropriate regulatory measures to be taken. Finally, the two regimes differ in their application of the precautionary approach. In the SPS Agreement, the burden of scientific proof rests on the country of import that wishes to impose a regulatory measure on a provisional basis (Article 5.7). For the Cartagena Protocol, the precautionary approach is the underlying rationale of the agreement itself. In this respect the regulatory rules and core aspects of the regimes are on a collision course and, as mentioned above, this is leading to a high degree of interfering interplay rather than synergy.

An additional problem relates to the non-ratification by a majority of GMO-exporting states of the Cartagena Protocol and, consequently, their non-involvement in the evolution of this regime. Though there is substantial overlap in membership of the two regimes, diffusive carriers of normative interplay between the two institutions seem to be lacking, in part because states with a major stake in the issue are pursuing their goals through separate institutional processes – the GMO exporters in the WTO and the European Union in the Cartagena Protocol. This impedes the ability of the institutions to achieve synergies from either normative or operational interplay. The present situation thus seems to be one of political interplay and "turf wars", as states seek to influence the institutional arrangement in which their vested interests seem most appropriately addressed. A general compatibility in provisions claimed by several commentators does not overcome the political driving forces in managing the interplay between the two regimes in the general direction of hindering their effectiveness, and rendering uncertain not only the effects of GMOs but also the effects of their international regulation. In the short term, the disagreements related to rules of precedence, the appropriate application of the precautionary approach and the nature of institutional overlap may impede the effectiveness of both institutions individually and of the global regulatory regime in general. It is not unlikely that the interplay of the Cartagena Protocol and the WTO will come before a WTO dispute settlement panel. The ruling in such a case may establish precedents that guide the future interactions of the regimes and delineate the relation between them. Nonetheless, as the regime constituted by the Cartagena Protocol evolves and debates regarding the precautionary approach and international standards for biosafety mature within the WTO and the Codex Alimentarius, there is still opportunity for international standards to develop and a growing consensus to arise in relation to the appropriate management of risk with regard to GMOs. However, as this chapter has demonstrated, perceptions of risk and precaution are not to be regarded merely as technical and scientific questions; they are political factors institutionalized in domestic and international regimes.

In section 2.2, three propositions were made regarding the political, normative and operational interplay of overlapping regimes. First, that core aspects and regulatory rules that constitute the defining features of a regime are sensitive to political conflict. Their relative compatibility therefore determines whether the political interplay of regimes is synergetic or impeding. Second, normative interplay is conditioned by the compatibility of the regimes and is more benign as regards the secondary aspects of a regime. Operational interplay – as an institutional adaptation to the above – is most likely in relation to programmatic regulations. This approach has a primary focus on ideational diffusion (normative interplay) and interest-driven politics (political interplay), reducing operational interplay to a secondary role. It thereby establishes an analytical hierarchy of causal mechanisms, whose strength is in understanding the dynamics of institutional interactions from an interest-driven perspective. Hence, if core values and regulatory rules are incompatible, one would expect institutional overlaps and interplay to have an impact on the effectiveness of the regimes. As institutions are negotiated and evolve within different issue areas and institutional environments, as is the case with trade agreements and MEAs, there is reason to believe that their core values, and frequently also their regulatory rules, will diverge. Matters of technical cooperation and the diffusion of knowledge through operational and/or normative interplay that might provide for synergetic interplay may in such cases be overshadowed by more fundamental political and institutional disputes.

A factor to note is that my analysis in this chapter has primarily taken the respective institutions and their overlaps at face value, that is, the texts of the constituting agreements. This omits the time factor in institutional interplay – how interplay may lead over time and through institutional practices to a more favourable context for synergetic interplay. At present, the issue area of biosafety and GMOs is in a political process of framing. The boundaries between trade and environmental aspects have not been drawn and the level of scientific and political uncertainty is high. As such, the conditions for synergetic interplay and regime effectiveness may seem particularly unfavourable. Whether the current situation prevails remains an open question and depends on the efforts of member countries to manage the situation.

Notes

This chapter was written by Are K. Sydnes while he was a postdoctoral fellow and associate professor in the Department of Political Science, University of Tromsø, Norway.

1. See Rosendal (2001a, 2001b).
2. For a discussion on this distinction, see Sydnes (2002).
3. Moreover, the Cartagena Protocol and the WTO agreements are also nested in broader institutional frameworks within their respective issue areas, that is, the CBD and the WTO as international organizations. Conflicts and synergies between the SPS Agreement and the Cartagena Protocol must therefore take into consideration their nested character.
4. This is partially implicit in Young's (1996) definition.
5. For example, it could be perceived that the negotiating parties to the Cartagena Protocol intentionally designed the agreement to circumvent the regulatory means of WTO agreements pertaining to trade in GMOs, by allowing for greater discretion to GMO-importing countries to regulate such trade. Similarly, initiatives at the WTO Ministerial Conference in Seattle to establish a framework for regulating trade in GMOs were regarded by many as an attempt to undermine the negotiation of the Cartagena Protocol by GMO-exporting countries (Falkner 2000: 305).
6. This is more common among nested regimes (Young 1996).
7. In her article on biodiversity, Rosendal (2001a) regards scientific uncertainty as a secondary aspect. In the case of biosafety, however, the treatment of scientific uncertainty by the Cartagena Protocol and the SPS Agreement may be regarded as a core aspect in the overlap of the regimes.
8. These categories deviate somewhat from the driving forces proposed by Stokke (2001).
9. The TBT Agreement explicitly delineates its relation to the SPS Agreement by stipulating that the provisions do not apply to sanitary and phytosanitary measures as defined under the SPS Agreement (TBT Agreement: Article 1.5).
10. It is worth noting that a similar formulation is to be found in the preamble of the WTO Marrakesh Agreement Decision in 1994 ("with the aim of making international trade and environmental policies mutually supportive"), later reiterated by the WTO members in the Doha Ministerial Declaration (WTO 2001: para. 31).
11. For a comprehensive discussion of the savings clause, see Safrin (2002).
12. For an extensive discussion on the methodological issues related to scientific uncertainty and risk assessments in the case of GMOs, see Busch et al. (2004).
13. Howse and Meltzer (2002: 492) argue that, in a case between parties to both the Cartagena Protocol and the WTO TBT Agreement, a WTO panel might find that the Cartagena Protocol would be considered a *lex specialis* to the TBT Agreement. Accordingly, measures under the Cartagena Protocol that on the basis of a risk assessment are deemed necessary to avoid adverse effects on biodiversity and/or human health should not be considered arbitrary or unnecessary. Moreover, they argue that Article 2.2 of the TBT Agreement, which allows for greater restrictiveness than the least-restrictive criteria in the face of potentially very serious or catastrophic effects, provides an opening for the application of the precautionary approach as dictated by the Cartagena Protocol (Articles 10.6/11.8). However, the extent to which the Cartagena Protocol would be considered as a *lex specialis* to the TBT Agreement, and thereby providing for normative interplay and operational interplay (delineation of jurisdiction), would be up to a WTO panel to consider. If this were the case, it would establish a precedent for further interactions between the Cartagena Protocol and the WTO agreements.
14. See the discussion in Busch et al. (2004).

REFERENCES

Andren, Robert and Bill Parish (2002), "Risk Assessment", in Christopher Bail, Robert Falkner and Helen Marquard (eds), *The Cartagena Protocol on Biosafety. Reconciling Trade in Biotechnology with Environment and Development?* London: RIIA/Earthscan, pp. 329–337.

Boutrif, Ezzeddine (2003), "The New Role of Codex Alimentarius in the Context of WTO/SPS Agreement", *Food Control* 14: 81–88.

Brack, Duncan and Kevin Gray (2003), "Multilateral Environmental Agreements and the WTO", Royal Institute of International Affairs, Sustainable Development Programme.

Busch, Lawrence, Robin Grove-White, Sheila Jasanoff, David Winickoff and Brian Wynne (2004), "Amicus Curiae Brief: Measures Affecting the Approval and Marketing of Biotech Products", Dispute Settlement Panel, World Trade Organization, April.

Cartagena Protocol (2000), *Cartagena Protocol on Biosafety to the Convention on Biological Diversity: Text and Annexes.* Montreal: Secretariat of the Convention on Biological Diversity; available at ⟨http://www.cbd.int/doc/legal/cartagena-protocol-en.pdf⟩ (accessed 5 July 2007).

Charnovitz, Steve (1999/2000), "The Supervision of Health and Biosafety Regulation by World Trade Rules", *Tulane Environmental Law Journal* 13: 271–302.

Cottier, Thomas (2002), "Implications for Trade Law and Policy: Towards Convergence and Integration", in Christopher Bail, Robert Falkner and Helen Marquard (eds), *The Cartagena Protocol on Biosafety. Reconciling Trade in Biotechnology with Environment and Development?* London: RIIA/Earthscan, pp. 467–481.

Falkner, Robert (2000), "Regulating Biotech Trade: The Cartagena Protocol on Biosafety", *International Affairs* 76(2): 299–313.

Gupta, Aarti (2000), "Governing Trade in Genetically Modified Organisms: The Cartagena Protocol on Biosafety", *Environment* 42(4): 23–33.

Gupta, Aarti (2001), "Advance Informed Agreement: A Shared Basis to Govern Trade in Genetically Modified Organisms?", *Indiana Journal of Global Legal Studies* (special edition on agricultural sustainability and genetically modified organisms) 9(1): 265–281.

Howse, Robert and Joshua Meltzer (2002), "The Significance of the Protocol for the WTO Dispute Settlement", in Christopher Bail, Robert Falkner and Helen Marquard (eds), *The Cartagena Protocol on Biosafety. Reconciling Trade in Biotechnology with Environment and Development?*, London: RIIA/Earthscan, pp. 482–496.

Kerr, William A. and Peter W. B. Phillips (2000), "Alternative Paradigms: The WTO versus the Biosafety Protocol for Trade in Genetically Modified Organisms", *Journal of World Trade* 34: 63–76.

Rivera-Torres, Olivette (2003), "The Biosafety Protocol and the WTO", *Boston College International & Comparative Law Review* 26: 263–323.

Rosendal, G. Kristin (2001a), "Impacts of Overlapping International Regimes: The Case of Biodiversity", *Global Governance* 7(1): 95–117.

Rosendal, G. Kristin (2001b), "Overlapping International Regimes. The Case of the Intergovernmental Forum on Forests (IFF) between Climate Change and Biodiversity", *International Environmental Agreements: Politics, Law and Economics* 1: 447–468.

Rosendal, G. Kristin (2005), "Governing GMOs in the EU: A Deviant Case of Environmental Policy-Making?", *Global Environmental Politics* 5(1): 82–104.

Safrin, Sabrina (2002), "Treaties in Collision? The Biosafety Protocol and the World Trade Organization Agreements", *American Journal of International Law* 96(3): 606–628.

SPS Agreement (1994), *The WTO Agreement on the Application of Sanitary and Phytosanitary Measures (SPS Agreement)*; text available at ⟨http://www.wto.org/English/docs_e/legal_e/15-sps.pdf⟩ (accessed 9 July 2007).

Stokke, Olav S. (1997), "Regimes as Governance Systems", in Oran R. Young (ed.), *Global Governance. Drawing Insights from the Environmental Experience*. Cambridge, MA: MIT Press.

Stokke, Olav S. (2000), "Managing Straddling Stocks: The Interplay of Global and Regional Regimes", *Ocean & Coastal Management* 43: 205–234.

Stokke, Olav S. (2001), "The Interplay of International Regimes: Putting Effectiveness Theory to Work", FNI Report No. 14/2001, Fridtjof Nansen Institute, Oslo.

Sydnes, Are K. (2001), "Regional Fishery Organizations: How and Why Organizational Diversity Matters", *Ocean Development and International Law* 34: 349–372.

Sydnes, Are K. (2002), "Institutional Interplay in International Fisheries Governance: The Evolution of the Role of Regional Fishery Organizations", doctoral dissertation, University of Tromsø, Norway.

Sydnes, Are K. (2005), "Regional Fisheries Organizations and International Fisheries Governance", in Syma A. Ebbin, Alf Håkon Hoel and Are K. Sydnes (eds), *A Sea Change: The Exclusive Economic Zone and Governance Institutions for Living Marine Resources*. Dordrecht: Springer Verlag, pp. 117–135.

TBT Agreement (1994), *Agreement on Technical Barriers to Trade. Annex to the Final Act Embodying the Results of the Uruguay Round of Multilateral Trade Negotiations*, Marrakesh, 15 April; text available at ⟨http://www.wto.org/English/docs_e/legal_e/17-tbt.pdf⟩ (accessed 9 July 2007).

Underdal, Arild (2004), "Methodological Challenges in the Study of Regime Effectiveness", in Arild Underdal and Oran R. Young (eds), *Regime Consequences: Methodological Challenges and Research Strategies*, Dordrecht: Kluwer, pp. 27–48.

WTO [World Trade Organization] (2001), *Doha WTO Ministerial 2001: Ministerial Declaration*, WT/MIN(01)/DEC/1, 20 November 2001; available at ⟨http://www.wto.org/English/thewto_e/minist_e/min01_e/mindecl_e.htm⟩ (accessed 9 July 2007).

Young, Oran R. (1996), "Institutional Linkages in International Society: Polar Perspectives", *Global Governance* 2: 1–24.

Young, Oran R. (2002), *The Institutional Dimensions of Environmental Change. Fit, Interplay and Scale*. Cambridge, MA: MIT Press.

Young, Oran R., with contributions from Arun Agrawal, Leslie A. King, Peter H. Sand, Arild Underdal and Merrilyn Wasson (1999), *Science Plan: Institutional Dimensions of Global Environmental Change*, IHDP Report No. 9. Bonn: International Human Dimensions Programme on Global Environmental Change.

5
Disentangling the interaction between the Cartagena Protocol and the World Trade Organization

Sebastian Oberthür and Thomas Gehring

1. Introduction

Despite growing interest by scholars and policy makers in the issue of institutional interaction, the conceptual development of the analysis of institutional interaction is still at an early stage. Institutional interaction (or interplay) generally refers to the phenomenon that international institutions influence each other in ways that are relevant for their development and effectiveness. A number of studies have highlighted the challenges and, less frequently, the opportunities that institutional interaction poses to international (environmental) governance (Young 2002).[1] Although a number of useful distinctions have been introduced (Young 1996, 2002; Young et al. 1999), no encompassing conceptual framework has emerged that could serve as a general basis for the empirical investigation of individual cases. In particular, few efforts have been made to clarify the causal mechanisms of how institutional interaction comes about (but see Stokke 2001a, 2001b).

It is therefore hardly surprising that studies on the relationship between the World Trade Organization (WTO) and the international biosafety regime based on the Convention on Biological Diversity (CBD) and its Cartagena Protocol on Biosafety have not yet been based on elaborate concepts of institutional interaction. The tense and potentially conflictual relationship between the international trade order represented by the WTO and various multilateral environmental agreements, such as the Cartagena Protocol, constitutes a particularly prominent element of the

Institutional interplay: Biosafety and trade, Young, Chambers, Kim and ten Have (eds), United Nations University Press, 2008, ISBN 978-92-808-1148-3

broader agenda of institutional interaction (e.g. Brack 2002; Schoenbaum 2002; Shaw and Schwartz 2002). Existing studies on the relationship between the Cartagena Protocol and the WTO have so far focused either on the importance of this relationship in the negotiations on the Cartagena Protocol or on the analysis of the scope for inconsistency and conflict between both sets of norms and rules.[2]

In this chapter, we offer an encompassing conceptual framework for analysing empirical cases of institutional interaction and their governance implications and illustrate its fruitfulness by applying it to the analysis of the relationship between the WTO and the Cartagena Protocol. The conceptual framework rests on two pillars. First, we suggest that the analysis of institutional interaction should best start from clearly identified *cases* of interaction involving two institutions connected by one causal relationship. Second, we introduce a number of distinct causal mechanisms and sub-types of these mechanisms that are characterized by discrete causal pathways and different governance conditions. We submit that every case of institutional interaction follows one of these causal mechanisms (section 2).

Applying this conceptual framework to the interaction between the international biosafety regime and the WTO regarding the regulation of international trade in genetically modified organisms (GMOs) reveals the story of a stepwise delimitation of the jurisdictions of the two institutions. The overall interaction consists of two separate instances in which inter-institutional influence runs in opposite directions. Both cases of interaction follow a common causal mechanism driven by the anticipated and actual commitments of parties under the two institutions. On the basis of the diverging objectives of the WTO and the biosafety regime as pursued by different groups of countries and policy communities, the interest of states parties to both agreements in avoiding inconsistent commitments drives both cases towards a jurisdictional delimitation of both institutions. Although each of the institutions involved has had a disruptive influence on the other side by restricting its room for regulatory activity, the resulting jurisdictional delimitation in which the Cartagena Protocol proved to show surprising strength has limited the potential for conflict between the two regimes (section 3).

The interest of states parties to both agreements in avoiding inconsistent commitments is also likely to mark the future evolution of the relationship between the regimes. *Ceteris paribus*, potential future political and judicial decision-making in the two regimes is constrained by existing commitments and both regimes will tend to be driven towards developing in compatible ways. Although major political initiatives to change the current jurisdictional balance hold little promise, the relationship between the two regimes will be worked out further in their implementation "on the ground" when states regulate trade in GMOs. Judicial decision-making on related challenges of national regulations under the WTO dispute settlement procedures or the compliance mechanism of the Cartagena Protocol may result in more delimitation of jurisdictions. Overall, there is a good chance that both regimes will develop in consistent ways in the future (section 4).

This analysis reveals institutional interaction as "jurisdictional delimitation" where two (or more) institutions compete for jurisdictional authority. This is a frequent occurrence in global environmental governance. It emerges that the structure of international governance provides for powerful forces driving institutions with differing objectives – such as the biosafety regime and the WTO – towards a jurisdictional balance that contains and limits the potential for conflict. Although the institution that first regulates the field acquires a first-mover advantage, which side is more successful in determining the eventual jurisdictional balance is subject to the vagaries of the political process based on interests of varying strength – with all the accompanying uncertainties and attractions. Thus, although the allegedly powerful WTO was first in structuring the regulatory field of international trade in GMOs, the seemingly weak Cartagena Protocol showed surprising strength in assuming regulatory authority and in exploiting the remaining space for rule-making.

2. Conceptualizing institutional interaction

2.1. Establishing a single cause–effect relationship between two institutions

Interaction, or interplay, between international institutions – be they international regimes or international organizations – requires that one institution (the source institution) affects the development or performance of another institution (the target institution) (Breitmeier 2000; Gehring and Oberthür 2004). To establish an instance of institutional interaction, we must identify:
1. the independent variable, namely the source institution and more specifically its particular component(s) or decision(s) from which influence originates;
2. the dependent variable, i.e. the target institution and more specifically its particular component(s) that are subject to influence originating from the source institution; and
3. a cause–effect relationship between the source institution and the target institution accounting for the identified effect.

Demonstrating a cause–effect relationship requires identifying the precise causal mechanisms that drive instances of institutional interaction and lead to an observable or anticipated effect within the target institution or the issue area governed by it (see section 2.2).

To allow for serious causal analysis, we suggest disaggregating complex real-world interaction situations into an appropriate number of *cases of interaction* with a single source institution, a single target institution and a unidirectional causal mechanism linking the two. Disaggregation will be especially necessary in three types of situation.

First, two institutions may be involved in numerous cases of interaction simultaneously, because they usually consist of varying components. For example, the Montreal Protocol on Substances that Deplete the Ozone Layer indirectly promotes the use of certain greenhouse gases (hydrofluorocarbons, HFCs) regulated under the Kyoto Protocol to the United Nations Framework Convention on Climate Change. Concurrently, it mandates the phase-out of chlorofluorocarbons (CFCs), which are potent greenhouse gases, thus supporting the objective of the international climate change regime. Moreover, the Montreal Protocol's non-compliance procedure provided a precedent for the elaboration of a similar component within the climate change regime (Oberthür 2001).

Second, an interaction situation may involve more than two institutions. For example, the Baltic Sea is affected by several global environmental regimes addressing, *inter alia*, oil pollution from ships and dumping of wastes at sea, an important regional regime (Helsinki Convention for the protection of the Baltic Sea), and overall arrangements such as the United Nations Convention on the Law of the Sea (Young 1996). It is likely that institutions co-governing this area interact with each other in various ways – either by affecting each others' performance or by influencing each others' decision-making processes.

Third, two or more institutions may "co-evolve" over time, with influence running back and forth between the institutions so that neither would exist in its current state in the absence of the other. In this case, distinguishing a suitable number of pairs of institutions connected by unidirectional causal pathways requires that we disaggregate the process analytically into sequential cases over time (Archer 1985; Carlsnæs 1992). For example, the co-evolution of the global Basel Convention on the transboundary movement of hazardous wastes and several related regional regimes can be disaggregated into two phases. In the first, analytical phase, the unsatisfying weakness of the global Basel Convention caused various developing countries to adopt separate regional regimes prohibiting the import of hazardous wastes. In the second phase, the existence of these regional regimes strengthened the hand of those advocating a ban on waste exports from the developed countries of the Organisation for Economic Co-operation and Development (OECD) to non-OECD developing countries; a ban was eventually agreed upon under the global regime (Clapp 1994; Meinke 2002).

The effects or consequences of a specific case of institutional interaction may be beneficial, adverse or neutral for the target institution. The main effects of institutional interaction occur in the target institution and can be assessed against this institution's prime objective. This builds on established research on the effectiveness of international institutions, which has also used the prime objective of the institution in whose domain the investigated effects occurred as the yardstick for assessing these consequences (e.g. Haas et al. 1993; Young 1999; Miles et al. 2002). If the effects of a case of institutional interaction support the objectives of the target institution, they create *synergy* between the two institutions involved. If they contradict the target's objective, they result in *disruption* and *conflict*. The aforementioned influence of the Montreal Protocol on the climate change regime provides suitable examples of both synergistic (CFCs) and disruptive effects (HFCs). The effects of an interaction may also be *indeterminate or neutral*, if they do not clearly hamper or reinforce the target institution's pursuit of its objective (Oberthür and Gehring 2006a).

Our research strategy is based on the assumption that complex interaction situations can be properly understood by disaggregating them into a given number of clear-cut cases. Although this approach allows for a clear causal analysis, it does not preclude the recombining of cases into a more complex picture. Co-evolution processes such as that between the Basel Convention and several regional regimes can be analysed and understood as a causal chain in which one case of interaction triggers the next. In other cases, clusters of parallel cases of interaction, such as those related to the Baltic Sea, may be recombined. Recombining cases of interaction into more complex interaction situations in principle also allows us to grasp the "emergent properties" of the larger situation, i.e. any logic or rationale that emerges only from the complexity of the specific situation or from the combination of cases (as opposed to the sum of the individual cases) (Gehring and Oberthür 2006).

2.2. *Causal mechanisms of institutional interaction*

To establish the cause–effect relationship between the source and the target institutions, we must identify the precise mechanism that drives an instance of institutional interaction (Elster 1989: 3–10; Hedström and Swedberg 1998). A causal mechanism is a set of statements that are logically connected and provide a plausible account of how a given cause leads to an observed effect (Schelling 1998). Since international institutions do not act on their own, actors such as states and other stakeholders that negotiate, develop and implement the institutions involved are essential elements of a causal mechanism that drives institutional interac-

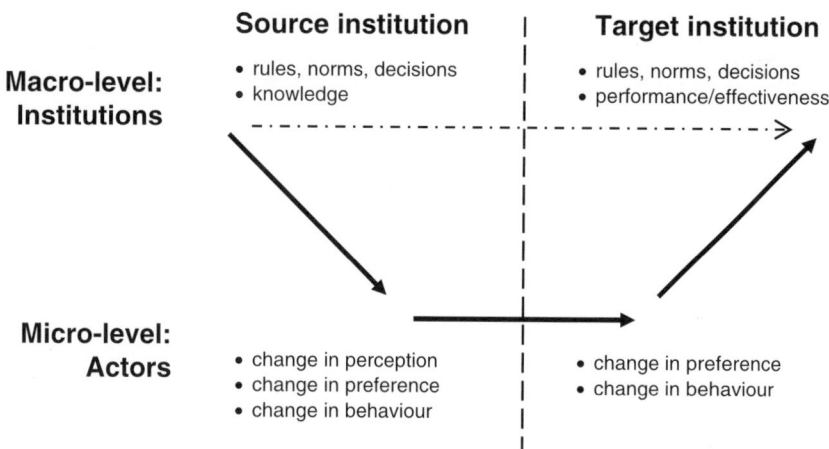

Figure 5.1 A causal mechanism of institutional interaction
Sources: Adapted from Coleman (1990: 1–23) and Hedström and Swedberg (1998: 21–23).

tion (see also Selin and VanDeveer 2003). To identify a causal mechanism driving institutional interaction, we must demonstrate (1) how the identified component of the source institution affects the decision situation of relevant actors, (2) that this leads to a change in the preferences or individual behaviour of actors relevant to the target institution, and (3) that these changes in preferences or behaviour produce the effect observed within the target institution or its issue area (see Figure 5.1).

In the following, we introduce four general causal mechanisms that may drive interaction between two international institutions. In two of these causal mechanisms – derived from different theories of institutions, negotiation theory and cooperation theory – the source institution directly influences the rule-making process of the target (section 2.2.1). The other two causal mechanisms are characterized by the fact that the source institution influences the implementation and effectiveness of the target institution (rather than its decision-making process) (section 2.2.2).[3]

2.2.1. Interaction influencing the decision-making process of the target institution

To influence the rule-making of another institution, the source institution has to influence the preferences of decision makers operating within the target institution. This may happen in one of two ways. First, the source institution may produce important new information, knowledge or ideas (cognitive interaction). Second, the commitments introduced by one institution may affect the preferences of actors negotiating within another institution. In addition, different sub-types of these causal mechanisms can be distinguished that vary with respect to key characteristics that matter from a governance perspective.[4]

Cognitive interaction is based solely on persuasion and may be conceived of as a particular form of inter-institutional learning (see also Stokke 2001b: 10). It occurs if information, knowledge or ideas (Risse-Kappen 1994; Yee 1996) produced within one institution modify the perception of relevant decision makers in another institution. It is based on the assumption that in real-world situations the rationality of actors is usually "bounded", either because the actors do not have all relevant information or because their information-processing capacity is limited (Simon 1972). Actors will therefore be prepared to adapt their perceptions to new information. These perceptions then shape their interests (Checkel 1998; Risse 2000). For this to happen, the collective decision-making process of the source institution must produce some new information that, upon its transfer to the target institution, changes the order of preferences of relevant actors and thus influences the collective negotiation process of the target institution and its output. The source institution does not exert any pressure. However, once relevant actors adapt their preferences, the consequences will be felt even by those participants in the process that have not been convinced.

We can distinguish between two ideal types of cognitive interaction depending on whether the interaction is initiated by the source or by the target. First, the target institution may initiate the interaction by drawing on aspects of other institutions as a *policy model* to devise a solution to a problem it faces. For example, negotiators of the Kyoto Protocol on climate change used the compliance system under the Montreal Protocol for the protection of the ozone layer as a policy model when elaborating a similar system (Oberthür and Ott 1999: 215–222; Werksman 2005). Second, the source institution may trigger the interaction by issuing a *request for assistance* to the target without having a particular means available to support its wish. For example, the Convention on International Trade in Endangered Species of Wild Fauna and Flora requested assistance from the World Customs Organization and Interpol in enforcing its trade restrictions (Lanchbery 2006).

Interaction through commitment is based on the fact that a commitment entered into within an international institution may change actors' preferences within another institution. It is closely related to Stokke's categories of "normative" and "utilitarian" interplay (Stokke 2001b). At a minimum, actors may be expected to develop an interest in passing compatible decisions in international forums to which they are party in order to be able to comply with their commitments. Otherwise, they could neither maintain a reputation of keeping their promises, providing the basis for future cooperation (Keohane 1984: 105–106; Young 1992: 175–176), nor preserve the benefits of cooperation that depend on stabilizing the prospects for compliance (Martin 1993). Members of an institution may also readily accept a commitment that they have already subscribed to in another institution because it does not involve additional costs. They may become interested in the transfer of a commitment to other institutions if this promises additional benefits such as the extension of the commitment to potential competitors. Being aware of the binding force of obligations, they may even gain an interest in adopting commitments in one institution in order to frame the policy choices available in another institution. As a result, actors may appear to pursue either "legal consistency" or "strategic inconsistency" (Raustiala and Victor 2004: 300–302).

Interaction through commitment involves, first, that members of the source institution agree upon an obligation that might be relevant for the target institution. Second, this obligation must actually commit one or more members of the source institution. Third, it must induce at least one member of the target institution to change its preferences. Fourth, this must influence the collective decision-making process of the target institution and its output. In the ideal case of interaction through commitment, an obligation originating from the source institution will affect a subsequent decision-making process of the target institution on a related subject. However, anticipated commitments to be entered into within the source institution may also trigger the mechanism.

Interaction through commitment requires a certain overlap of both the memberships and the issue areas of the interacting institutions and, in contrast to cognitive interaction, it does not completely depend on the target institution. Without a jurisdictional overlap of issue areas, commitments entered into in one institution could not become relevant to another. Without overlapping memberships, the target institution would remain unaffected because none of its members would be subject to relevant commitments. Whereas in the case of cognitive interaction the relevant actors modify their preferences entirely voluntarily, interaction through commitment incites them to do so because of the costs and benefits involved.

We can distinguish three ideal types of interaction through commitment:

1. *Interaction between nested institutions* may help extend an obligation from a smaller institution to an institution with a larger membership and similar objectives. Because all members of the smaller institution will tend to favour the extension of the obligation, the relevant coalition in the bigger institution will be strengthened. For example, the ban on the dumping and incineration of waste at sea within the regional regime for the protection of the North-East Atlantic (the Oslo and

Paris Conventions, or OSPAR) led to the adoption of similar measures within the global London Dumping Convention (Meinke 2002; Skjærseth 2006).
2. A target institution may take on obligations from a source institution because the memberships of both institutions are largely identical, so that members of the target can hardly object to such a transfer. Such interaction will be relevant only if the target institution has *additional means* available to foster compliance with and enforcement of the obligation. For example, once political agreement had been reached within the soft-law-based International North Sea Conferences in the 1980s and 1990s, parties to OSPAR easily agreed on hard-law targets for reducing pollution (Skjærseth 2006). Both policy diffusion between nested institutions and interaction resulting in the activation of additional means will usually enhance global governance.
3. In contrast, institutions with differing objectives will tend to regulate overlapping issues in diverging ways. Because such cases entail the danger that states members of both institutions become subject to incompatible commitments, they create a demand for *jurisdictional delimitation*. They may be resolved amicably, but can easily lead to political conflict if different actors involved in the decision-making process prefer differing allocations of jurisdictional authority. The tense relationship between the WTO and the Cartagena Protocol (and other multilateral environmental agreements) provides a case in point. Section 3 will therefore shed more light on this type of institutional interaction through commitment.

2.2.2. *Interaction directly affecting the implementation and effectiveness of the target institution*

In addition to the decision-making of the target institution, the source institution may directly affect the target's implementation and effectiveness. To derive suitable causal mechanisms we build on the distinction between the output (i.e. rules and norms), the outcome (i.e. the effects on the behaviour of relevant actors in the issue area) and the impact of an institution (i.e. its effect on the environment or other ultimate target of governance) established in research on the effectiveness of international institutions (Underdal 2004). Accordingly, a distinction can be made between behavioural interaction at the outcome level and impact-level interaction.

Behavioural interaction exists if an international institution induces behavioural changes within the issue area governed by another and thereby influences its performance. All international governance institutions are designed to influence the behaviour of relevant actors in order to achieve their objectives, such as protecting the environment or liberalizing international trade (Young 1992; Levy et al. 1995). Such behavioural effects may also directly or indirectly affect the implementation of another institution. For example, increased use of HFCs resulting from the Montreal Protocol is immediately relevant for the implementation of the climate change regime, which aims at reducing the emissions of greenhouse gases, including HFCs (Oberthür 2001).

Impact-level interaction occurs if an institution's impact on its ultimate target of protection, such as free trade or protection of the ozone layer, affects the ultimate target of the other institution. In contrast to the other causal mechanisms, impact-level interaction does not involve social interaction between the two interacting institutions, but is based on a scientific link between the two targets of governance involved. It may therefore be considered of lesser interest to social scientists.

A stylized example that we owe to Arild Underdal may illustrate this least-intuitive causal mechanism. Consider that the ultimate targets of two separate international institutions are protection of the stocks of cod and protection of the stocks of herring. Because cod eat herring, successful protection of cod, resulting in a growing population of this species, will unintentionally decrease the population of herring. In this case, the two institutions are not linked at the level of output (neither the norms of the cod regime nor the knowledge produced within it influence the norms protecting herring) or through behavioural changes (decreased fishing of cod does not directly influence the fishing activities related to herring). They are "functionally linked" (Young et al. 1999; Young 2002) at the impact level because the effects of the source institution on its ultimate regulatory target (the cod population) affect the ultimate regulatory target of the target institution (the herring population).

Both behavioural interaction and impact-level interaction are characterized by a high ability of the source institution to influence the target unilaterally. In contrast to interaction at the output level, these types of interaction do not depend on a decision within the target institution. A collective decision by the target institution or by the source institution (or a "political linkage" between them – Young et al. 1999: 50) in response to the effects of behavioural or impact-level interaction is possible, but such interaction "management" (Stokke 2001a) is not an essential element of these causal mechanisms and the effect would also occur without a policy response.

2.3. Summary

Our conceptual approach enables us to engage in an exact and differentiated diagnosis of institutional interaction, which is a necessary precondition for devising adequate policy responses. The conceptual foundation results in two-fold guidance for analysing the interaction between the WTO and the Cartagena Protocol.

First, the interrelationship between the WTO and the Cartagena Protocol should be disaggregated into clearly identifiable cases of interaction. Instead of determining whether the two instruments are compatible or have a potential for conflict, our approach leads us to ask how exactly each side has influenced the other, in what ways and with what consequences. It also allows us, by means of the recombination of related cases, to obtain an analytically clear overall picture of the interaction situation and its effects.

Second, particular attention has to be paid to identifying the causal mechanisms and ideal types of institutional interaction that each case of interaction between the WTO and the Cartagena Protocol follows. Each of the ideal types and causal mechanisms is characterized by different rationales, driving forces and key governance conditions, resulting in varying effects/consequences. For example, no other means than persuasion is available to the source institution in cases of cognitive interaction, whereas it can and does influence the costs and benefits accruing to certain policy options of the target institution in cases of interaction through commitment. Within the realm of interaction through commitment, there is a large range of differences distinguishing interaction among nested institutions, which tend to enhance global governance by means of horizontal policy diffusion, from issues of jurisdictional delimitation, which tend to lead to disruption and conflict. As a consequence, different causal mechanisms and ideal types require different responses and provide different political opportunities. Hence, our conceptual approach to analysing institutional interaction promises to deliver results that are relevant not only for scholars but also for policy makers.

3. Interactions between the WTO and the biosafety regime

Applying our conceptual framework to the interaction between the biosafety regime and the WTO reveals that both sides have influenced each other in the past. Disaggregating the interaction situation, we first analyse the influence of the biosafety regime on the WTO (section 3.1) before turning to the ways in which the WTO has affected the Cartagena Protocol (section 3.2). Overall, these disruptive cases of interaction have resulted in a stepwise delimitation of jurisdictions between the two institutions (section 3.3).

3.1. *The biosafety regime's influence on the WTO: Assuming regulatory authority*

Throughout the 1980s, both economic/trade interests and environmental interests made first attempts to occupy the newly emerging regulatory field of international trade in genetically modified organisms (GMOs). As pointed out in the introductory chapter to this volume, the Organisation for Economic Co-operation and Development, the United Nations Industrial Development Organization, the World Health Organization, the United Nations Food and Agriculture Organization, as well as the United Nations Environment Programme all became involved in the field (Pythoud and Thomas 2002: 40; Zedan 2002: 28–33).

Two aspects of international trade in GMOs were in potential conflict, virtually precluding joint regulation of the subject matter. First, biotechnology created demand for the establishment of an international market so that GMOs could be traded like other goods. This demand was in line with the objectives of the world trade regime – then primarily built upon the General Agreement on Tariffs and Trade (GATT) – and was reflected in the trade interests of GMO exporters. At the same time, the spread of GMOs entailed new risks for the environment and for existing socio-economic structures. These risks created demand for regulation to protect "biosafety" – here understood to mean the protection of biological diversity and established socio-economic structures against the risks associated with the spread of GMOs. Protective regulation would restrict markets, so that the two demands for market creation and market restriction were pointing in opposite directions (on the issue in general see e.g. Falkner 2000: 300–303).

Trade interests were first to occupy the newly emerging ground of regulating trade in GMOs through the WTO agreements of 1994. GATT does not limit the general ability of countries to restrict trade in GMOs (Rivera-Torres 2003: 289–291; Boisson de Chazournes and Mbengue 2004: 291–294). Consequently, members of the world trade regime prior to the establishment of the WTO agreements were free to restrict the market access of genetically modified products, as long as this restriction applied to GMOs of both foreign and domestic origin (the principle of national treatment). The new Agreement on the Application of Sanitary and Phytosanitary Measures of 1994 (SPS Agreement) changed this situation and made import restrictions subject to a number of requirements. In particular, it requires that measures restricting the import of GMOs (and other products) for sanitary or phytosanitary reasons be based on sufficient scientific evidence and a risk assessment that conforms to certain standards defined by the agreement and further developed through interpretation by the WTO dispute settlement organs (e.g. Rivera-Torres 2003: 296–298). Socio-economic considerations, although not explicitly excluded, are not recognized in this risk assessment (Gupta 2001). Furthermore, the SPS Agreement subjects precautionary measures to a number of conditions. In particular, it requires that scientific evidence be insufficient; that measures be adopted "on the basis of available pertinent information"; that the party concerned seeks to obtain the information necessary for a full risk assessment; and that it reviews the measure "within a reasonable period of time" (SPS Agreement 1994: Article 5.7). The WTO Agreement on Technical Barriers to Trade (TBT), which was strengthened in 1994, may be of lesser relevance for trade in GMOs. It establishes certain criteria that technical regulations such as labelling requirements have to fulfil and might apply to restrictions on GMO imports for other than sanitary or phytosanitary purposes.[5]

Against this backdrop, the biosafety regime was first influenced by the WTO when it claimed authority for regulating trade in GMOs in 1995. In 1995, parties to the CBD launched negotiations on a biosafety protocol on the basis of Article 19.3 of the CBD, which had envisaged the elaboration of such a protocol (Falkner 2002: 6). To be sure, the undertaking of establishing a biosafety protocol was broader than assuming authority for trade in GMOs – or, as they are called in the context of the biosafety regime, "living modified organisms" (LMOs) – from the WTO. Most importantly, the protocol aims at providing guidance and assistance to developing countries that lack sufficient capacities for enacting and implementing suitable domestic rules. However, defining more clearly the *rights* of countries to restrict GMO imports – as opposed to the restrictions of these rights under the WTO – formed a central part of the endeavour (Falkner 2000: 302–303). In this respect, the Cartagena Protocol was to elaborate more specific rules for one sub-area of international trade regulated by the WTO, namely trade in GMOs.

The launching of negotiations on a biosafety protocol under the CBD influenced the international interest constellation regarding the regulation of trade in GMOs. With the negotiating mandate of 1995, parties to the CBD had in effect committed themselves to introducing specific rules on restricting trade in GMOs under the CBD. This commitment significantly strengthened those in the WTO interested in preventing the WTO from further regulating trade in GMOs. For those members of both regimes that were in favour of free trade of GMOs, negotiating and introducing relevant rules in the WTO would have meant not honouring the commitment made in the negotiating mandate under the CBD. For those countries advocating regulation under the CBD, it would have meant weakening the jurisdictional authority of the biosafety regime. For both sides, it would have entailed the danger of elaborating inconsistent provisions.

As a result of this influence on actors' preferences, the biosafety regime helped block off attempts to further regulate biotechnology under the WTO in 1999. Proposals for regulating biotechnology under the WTO were made by Canada, Japan and the United States in the run-up to the WTO Ministerial Conference in Seattle in 1999. Given the failure to reach agreement under the CBD earlier that year, trade interests saw a window of opportunity for reclaiming regulatory authority. The proposals were successfully rejected, in particular by developing countries, with explicit reference to the ongoing negotiations under the CBD. They considered the CBD to be the preferable forum and feared that the WTO would seize exclusive jurisdiction over the issue (Falkner 2000: 305; Palmer et al. 2006).

This influence of the biosafety regime on the WTO followed the causal mechanism of interaction through commitment and displayed the characteristics of the ideal type of jurisdictional delimitation, with disruptive effects on the target institution. The case may not be easily recognized because it resulted in a non-regulation. On the basis of the diverging objectives of the two institutions, the CBD successfully claimed and assumed regulatory authority from the WTO for the issue of trade in "living" GMOs in 1995. This had a noticeable effect on the interests of the members of the biosafety regime, which were for the most part also parties to the WTO. As a result, the WTO – which covers a much broader area than GMOs – lost its ability to elaborate detailed rules for the sub-area of trade in GMOs, which was carved out from its jurisdiction and shifted to the biosafety regime. This disruptive effect on the WTO is evident from the failure to establish talks on biotechnology under the WTO in 1999. Overall, the case demonstrates a rather surprising strength of the seemingly weak biosafety regime vis-à-vis the supposedly much stronger WTO.

3.2. *The influence of the WTO on the design of the Cartagena Protocol*

Although the emerging biosafety regime had assumed regulatory authority from the WTO, the negotiators of the Cartagena Protocol exercised this authority "in the shadow of the WTO". Although it was the parties to the biosafety regime (and not the parties to the WTO) that decided on the restrictions on trade in GMOs, all industrialized countries and many important developing countries were also members of the WTO. Other developing countries such as China were expecting to join the WTO soon. These countries were influenced and limited in their choices by existing WTO rules. They had to take into account the possible implications of the emerging Cartagena Protocol for the interpretation of WTO law. They also had a strong incentive to avoid incompatibilities between the two instruments, because inconsistent rules could lead to costly conflicts in the implementation of both agreements.

"Interaction through commitment" of the jurisdictional delimitation ideal type disrupted the effectiveness of the Cartagena Protocol in two respects. First, the "shadow" of the WTO left its imprint, in particular on the rules on risk assessment and precaution (and socio-economic considerations). These rules are at the centre of the Cartagena Protocol because they specify the criteria that guide the transboundary movement of LMOs (Graff 2002; Palmer et al. 2006). Second, the influence of the WTO led to the elaboration of provisions clarifying the relationship of the Protocol with "other international agreements".

3.2.1. Risk assessment, precaution and socio-economic considerations

Existing WTO rules affected the preferences of the countries negotiating the Cartagena Protocol, in particular with respect to the provisions on risk assessment, precaution and socio-economic considerations. These provisions had a high potential for inconsistencies with WTO law, which could have led to unwelcome conflicts in the implementation of both agreements. Counterfactual reasoning reveals that considerations of consistency probably were least important for the position of the so-called Miami Group of agricultural exporting countries (including Argentina, Australia, Canada, Chile, the United States and Uruguay), which most forcefully used the argument of WTO compatibility. These countries would have strongly opposed restrictions on the free trade in GMOs/LMOs even in the absence of relevant WTO rules. However, the hands of the Miami Group were strengthened in the Cartagena negotiations because most other countries, including the proponents of strong biosafety provisions such as the European Union and the Like-Minded Group of developing countries, had an interest in avoiding incompatible regulations. Even though these countries vigorously defended the regulatory objectives of the biosafety regime, their stance was significantly softened owing to their wish to avoid incompatible commitments that could diminish the effectiveness of one or even both agreements. US non-membership in the CBD further strengthened this interest because it increased the danger of challenges under the WTO dispute settlement procedures (Palmer et al. 2006; on the negotiating process see Falkner 2000, 2002).

As a result, the risk assessment provisions of the Cartagena Protocol largely match and elaborate those of the WTO SPS Agreement. The Protocol obliges exporters of certain LMOs (such as LMOs for use as seeds) to seek and receive the so-called "advance informed agreement" (AIA) of the importing country before exporting. Articles 10 and 15 of the Protocol require importing countries to base their related decisions on a risk assessment and to follow certain procedural steps. Developing countries and formerly socialist countries with "an economy in transition" may also opt to subject LMOs "for direct use as food or feed, or for processing" to a risk assessment before taking a decision on import (Cartagena Protocol 2000: Article 11.6). LMOs for direct use as food or feed or for processing are otherwise exempted from the AIA procedure and account for about 90 per cent of trade in GMOs (Eggers and Mackenzie 2000: 525, 530). Article 15 lays down some criteria for conducting risk assessments that are further specified in Annex III of the Protocol. These criteria are more elaborate than those of the SPS Agreement but do not diverge from them in any meaningful way. In particular, both regimes clearly require risk assessment to be science based (e.g. Eggers and Mackenzie 2000: 539; Stoll 2000: 113–114; Rivera-Torres 2003: 296–298, 313–314).

The Protocol provisions on precaution and socio-economic considerations are more problematic, but can also be interpreted in ways that are consistent with WTO rules. According to Articles 10.6 and 11.8 of the Protocol, "lack of scientific certainty due to insufficient relevant scientific information and knowledge regarding the extent of the potential adverse effects of a living modified organism on the conservation and sustainable use of biological diversity in the Party of import, taking also into account risks to human health, shall not prevent that Party from taking a decision, as appropriate, with regard to the import ..., in order to avoid or minimize such potential adverse effects". In contrast to Article 5.7 of the SPS Agreement, this language does not require precautionary action under the Cartagena Protocol to be based upon "available pertinent information" or to be adopted "provisionally". Nor does it require parties to seek to obtain additional information and to review the precautionary measure "within a reasonable period of time" (Stoll 2000: 114–117). Furthermore, Article 26.1 of the Protocol goes beyond the SPS Agreement by allowing countries to "take into account, consistent with their international obligations, socio-economic considerations arising from the impact of living modified organisms on the conservation and sustainable use of biological diversity". These provisions of the Cartagena Protocol on precaution and socio-economic considerations differ from the existing WTO rules, but are *a priori* not inconsistent with them. Both sets of rules *can* be interpreted in consistent ways (e.g. Eggers and Mackenzie 2000: 539–540; Gupta 2001: 277–279; Howse and Meltzer 2002: 488–491; Rivera-Torres 2003: 308, 314–315; Boisson de Chazournes and Mbengue 2004: 295–297). At the same time, the Cartagena provisions may be employed in the interpretation of the related WTO rules, which may result in more leeway being granted to biosafety interests. It is unclear how and to what extent exactly this might be done by the WTO dispute settlement bodies (Eggers and Mackenzie 2000: 541–542; Howse and Meltzer 2002: 488–491; Boisson de Chazournes and Mbengue 2004: 297–301).

Overall, the provisions of the Cartagena Protocol relating to risk assessment, precaution and socio-economic considerations significantly limit the potential for conflict between the Cartagena Protocol and the SPS Agreement. In particular, no obvious incompatibilities exist, so that countries do not face the choice between the Scylla of not complying with the SPS Agreement and the Charybdis of not fulfilling their obligations under the Cartagena Protocol. Both instruments can be interpreted in mutually supportive ways. At the same time, there is no certainty that the two agreements *will* be interpreted in compatible ways. Since the area of compatible interpretations has loose boundaries, different actors may arrive at different interpretations and may wish to exploit the room for manoeuvre that exists in this respect.[6] Therefore, a limited potential remains for interpreting in varying ways what may be considered "compatible" or "mutually supportive" (see e.g. Eggers and Mackenzie 2000; Stoll 2000; Gupta 2001; Cottier 2002; Howse and Meltzer 2002; Safrin 2002b; Rivera-Torres 2003).

3.2.2. The relationship of the Protocol with other international agreements

The significance of clarifying the relationship between the WTO and the Cartagena Protocol is rooted in the remaining room for different interpretations of the two instruments by individual parties implementing them, by the Conferences of Parties and by the relevant judicial processes overseeing their implementation. In particular, the powerful dispute settlement bodies of the WTO are mandated to take into account general rules of international law and have developed a record of doing so, including paying attention to the provisions of multilateral environmental agreements (e.g. Howse and Meltzer 2002; Boisson de Chazournes and Mbengue 2004: 297–301).

On this basis, one of the main contentious issues in the biosafety negotiations concerned the relationship between the future Cartagena Protocol and the WTO. How the balance between the two regimes should be defined was primarily responsible for the one-year delay in the adoption of the Cartagena Protocol. Negotiators of the Cartagena Protocol were aware that the two agreements could not be considered in isolation. They knew and anticipated that the Cartagena Protocol had the potential of influencing their obligations under the WTO. It may therefore not be surprising that it was the main trading blocs, namely the United States and the European Union, that became particularly involved in the debate. Their WTO obligations not only influenced their interests but were even constitutive of them. The Miami Group of LMO exporters in particular feared that, without clarification, the customary *lex posterior* rule might have suggested that the later Cartagena Protocol takes precedence over earlier WTO rules (Safrin 2002a: 439). They therefore suggested a "savings clause" that would have in effect subordinated the Protocol to other international agreements, including the WTO. The European Union in particular was opposed to such subordination and favoured a balance that would tend more towards the Cartagena Protocol (while paying due respect to its obligations under the WTO). In fact, it would have been content with not addressing the relationship with other international organizations at all (Safrin 2002a; Alfonso 2002).

The resolution of this issue was found in a compromise somewhere between the two positions. The balance between the Cartagena Protocol and the WTO was defined in a way that does not subordinate either side to the other, but treats them as equals. The Protocol parties express their willingness to interpret the Protocol and the WTO agreements in mutually supportive and compatible ways and, in effect, present this as a "peace offer" to be reciprocated by the WTO. To this end, three elements were incorporated in the preamble of the Protocol. First, parties to the Protocol recognized "that trade and environment agreements should be mutually supportive with a view to achieving sustainable development". This language implies the suggestion or request to the WTO to join the Cartagena Protocol in interpreting its own provisions in ways that would be compatible with the Protocol obligations. Second, the Protocol parties emphasized that the Protocol "shall not be interpreted as implying a change in the rights and obligations of a Party under any existing international agreements". They thus offer to limit their own regulatory and judicial authority by promising that the Protocol organs would not interpret the Protocol in ways that would be incompatible with WTO obligations. Third, parties expressed their understanding "that the above recital is not intended to subordinate this Protocol to other international agreements". This provision reinforces the Protocol parties' claim for authority to take relevant decisions and rejects a notion that the Protocol would be subordinate to the WTO. Overall, the preamble tries to prevent conflicts between the two regimes by keeping a fine balance between limiting and upholding the Protocol's own regulatory authority, while offering guidelines for a peaceful coexistence with the WTO (for analyses, see Safrin 2002a; Cottier 2002; Howse and Meltzer 2002).

3.3. Overall assessment: Stepwise jurisdictional delimitation

The interaction between the biosafety regime and the WTO involves at least two cases in which influence runs in opposite directions. The biosafety regime exerted noticeable influence on the WTO when it claimed authority to regulate international trade in GMOs/LMOs in the mid-1990s because this move essentially excluded further regulation of this area by the WTO. The pre-existing regulations of the WTO SPS Agreement, in turn, heavily influenced several elements of the Cartagena Protocol, including its preamble and the provisions on risk assessment and precaution. Identifying these cases and their causal pathways requires disaggregating the interaction relationship between the two institutions.

The competitive quest of the WTO and the biosafety regime for regulatory authority over the newly emerging issue of international trade in GMOs has resulted in a stepwise delimitation of jurisdictions. On the one side, the biosafety regime's launching of negotiations on a biosafety protocol in 1995 established its regulatory authority over trade in LMOs and thereby removed it, to a large extent, from the WTO. On the other side, the WTO agreements of 1994 severely limited the options available to biosafety negotiators for regulating trade in GMOs, thereby casting their "shadow" over the emerging biosafety regime. Although each case of interaction thus had a disruptive effect on the respective target and involved serious political conflicts between the respective policy communities, from a broader governance perspective the regulatory competition of the two regimes has led to a far-reaching jurisdictional delimitation. With the subsequent limited room for incompatible interpretations and implementation, the potential for conflict has been greatly reduced (even though the result may not be to the liking of one or the other side). Both institutions were thus driven towards jurisdictional delimitation even without a related overarching institutional structure governing this process.

These cases of inter-institutional influence followed the causal mechanism of interaction through commitment and, more specifically, the ideal type of jurisdictional delimitation. The interaction was premised on a significant overlap in membership of the WTO and the biodiversity regime. As a result, several countries that were members of both institutions were committed under one institution when negotiating within the other institution. Incompatible commitments were looming because both the Cartagena Protocol and the WTO aim at regulating international trade in GMOs/LMOs and thus overlap in their jurisdictional scope. In doing so, they pursue different logics and the opposite objectives of free international trade (WTO) and of biosafety (Cartagena Protocol). Consequently, importing and exporting countries and the related policy communities have different preferences as to the institutional home of regulation. At the same time, regulation by one institution inevitably affects the chances of the other side to realize its objectives. Based on the interest of members of both institutions in avoiding incompatible commitments, this situation created a demand for delimitation of the jurisdictions of the two institutions.

The interaction between the biosafety regime and the WTO reveals the structure of jurisdictional delimitation issues in global environmental governance. Members of both the source institution and the target institution are in a "mixed motive" situation that resembles the game-theoretic constellation of the Battle of the Sexes (Stein 1982; Keohane 1984). On the one hand, they possess a common interest in some sort of separation of jurisdictions because neither the WTO nor the Cartagena Protocol can be satisfied with a conflict that disrupts both international trade and environmental protection. On the other hand, the constituencies of the two institutions have conflicting preferences that make it notoriously difficult to find a mutually acceptable solution. Actors favouring free trade will advocate regulation by the WTO, whereas countries (and other actors) struggling for far-reaching domestic environmental regulation will prefer enlarged jurisdiction of the Cartagena Protocol. The governance challenge consists in arriving at a delimitation of jurisdictions that balances the diverging interests and realizes the common interests.

In jurisdictional delimitation cases, the institution that regulates first possesses a "first-mover advantage" (Héritier 1996; Mattli 2003). "Battle of the Sexes" equilibria are comparatively stable. Commitments existing within one institution will therefore almost automatically limit the room for manoeuvre in negotiations within the other institution (if conflict is to be avoided). With the conclusion of the WTO agreements in 1994, the world trade regime thus secured a first-mover advantage by determining the requirements that restrictions by importing countries had to meet. Although some members of the biosafety regime might have preferred open conflict with the WTO, member states of both institutions had an interest in avoiding incompatible commitments and open conflict.

Therefore, the Cartagena Protocol's successful assumption of jurisdictional authority from the WTO came at a price: it was dependent on accepting the WTO's basic regulatory approach. As a consequence, the Cartagena Protocol comes close to setting *maximum* standards, which is unusual for international environmental agreements. Most of them define *minimum* levels of action that countries are required to take because countries have incentives to implement low protection standards; exceeding these standards is unproblematic and even contributes to achieving the environmental objective pursued. In contrast, the Cartagena Protocol does not just determine a (minimum) standard to be followed by all countries (including developing countries). By introducing certain criteria for risk assessment and precaution, it also in effect limits the level of protection that countries can justify regarding the import of LMOs. To be sure, the Protocol stops short of requiring importing countries to permit LMO shipments into their territory if the prescribed risk assessment does not, with sufficient scientific certainty, identify a certain level of risk – even though one could argue that such a requirement may be implicit in its rules. In effect, however, it specifies the pre-existing WTO maximum standards of regulation that WTO members are required not to exceed in order to facilitate free trade. Only by accepting the market creation logic and the existing limitations established by the WTO could the Protocol successfully determine what it deemed to be consistent with pre-existing WTO rules. As a result, the Protocol primarily further specifies, interprets and develops pre-existing WTO rules with respect to GMOs/LMOs, exploiting the room for interpretation that the relevant WTO agreements had left. Even if not in a formal legal sense (Howse and Meltzer 2002), the Protocol de facto constitutes a *lex specialis* to the WTO agreements (in particular the SPS Agreement).

4. Possible future interaction and policy implications

Even after the delimitation of jurisdictions of the WTO and the Cartagena Protocol, interaction between the two institutions can be expected to continue. General rules can never account for all specific circumstances of the particular cases to which they apply. They must be interpreted implicitly or explicitly, and interpretation provides margins of discretion. Assuming that both the relevant WTO rules, especially its SPS Agreement, and the Cartagena Protocol remain unchanged for the foreseeable future, several possible future cases of interaction can be anticipated. In this section, we spell out these possible future interactions and explore both their origins and effects on the basis of a *ceteris paribus* assumption: we assume that countries will continue to differ in their interests as to the appropriate balance between trade and biosafety objectives, but that the contracting parties will be prepared to honour clear-cut commitments entered into under either institution.[7]

4.1. Future behavioural interaction

Interaction between the two institutions will inevitably continue because the implementation of their rules by the contracting parties is closely interdependent. Through their unilaterally determined domestic action, the member states will inevitably influence the trade-off between the competing objectives of free trade and of biosafety and will affect the performance of both the relevant international institutions. Every state action concerning trade in GMOs/LMOs simultaneously implements the rules of the SPS Agreement and the Cartagena Protocol. The more a country restricts the import of GMOs/LMOs under the Cartagena Protocol, the more it will undermine the free trade objective of the WTO. And the more liberally a state regulates such imports in line with the free trade objective, the less it will conform to the objective of the Cartagena Protocol to ensure an adequate level of protection against the risks associated with trade in GMOs/LMOs (see also Burgiel 2002: 59–60).

This interdependence reflects the logic of behavioural interaction (see section 2.2.2). It originates not immediately from the rules of the source institution but from the implementation of these rules by a single actor operating within the institution's issue area. The influence is directed not at the decision-making of the target institution but at its performance within its own issue area. The occurrence of behavioural interaction does not require collective decisions but results from the unilateral implementation of existing decisions by relevant public or private actors. Over time, the implementation of individual actors will generate an order reflecting the accepted balance between trade interests and environmental interests. The exact nature of this order is not yet known because implementation of the biosafety regime is still at an early stage (see Gupta, Chapter 2 in this volume).

Ceteris paribus, biosafety interests tend to have an advantage over free-trade interests in behavioural interaction because they have better opportunities to influence the balance between the two objectives in their national implementation. A country predominantly interested in biosafety can expect to realize this objective in its implementation, because it controls its own domestic customs boundaries. Because of the interdependence of trade and biosafety objectives, this will undermine the free trade in GMOs/LMOs. In contrast, a country prioritizing free trade cannot achieve its objective unilaterally, because at least two countries are involved in international trade. Thus, exporting countries are not in control of the import side. WTO law also prohibits unilateral trade sanctions. Overall, therefore, countries interested in biosafety are in a better position to move the balance between the two objectives in their preferred direction in domestic implementation than are countries favouring free trade. In accordance with our *ceteris paribus* assumption, this advantage reaches its limits where national implementation clearly disregards existing WTO commitments. We should also caution that obviously other factors will influence whether or not interested countries will actually exploit the opportunity for more restrictive regulation of GMO imports.

4.2. Possible further action within the WTO

Within the WTO, the delimitation of jurisdictions between the two institutions involved might be further defined in response to the emergence of undesired restrictive GMO import regulation. The WTO provides for at least two ways of diminishing the discretion left by the rules of the SPS Agreement and the Cartagena Protocol, namely the court-like dispute settlement mechanism and political decision-making by the contracting parties. A finding of the dispute settlement mechanism could result in a further definition of the applicable rules, because every (quasi-)judicial application of existing rules necessarily involves law-making (Shapiro 1981: 28–36).

The powerful WTO dispute settlement mechanism provides a promising forum for countries with strong trade interests, because the mechanism has the objective of protecting free trade against unjustified restrictions. Interested countries may therefore challenge bold unilateral restrictions on trade in GMOs/LMOs under the Cartagena Protocol. The scope for such legal challenges primarily stems from the remaining potential for tensions in the interpretation of the Cartagena Protocol and the WTO agreements by individual countries. Whereas the Cartagena Protocol provides additional justification for countries interested in restricting trade in GMOs/LMOs – and may be supported by provisions of the Codex Alimentarius Commission and the International Plant Protection Convention (Homeyer 2006) – countries privileging trade can employ judicial action to limit this opportunity. A first relevant US challenge of EU restrictions on trade in GMOs came before the WTO in 2003. Even though the challenge was not directly related to the Protocol, its findings may still result in the judicial development of rules relevant for the interaction between the WTO and the Cartagena Protocol (Boisson de Chazournes and Mbengue 2004; Gupta, Chapter 2 in this volume).

In particular, countries with strong biosafety interests could attempt to achieve political decisions within the WTO to shift the delimitation of jurisdictions in their favour. Any political WTO decision addressing the issue would have to at least acknowledge the Cartagena Protocol. First, a proposal is pending for the SPS Committee to recognize the Cartagena Protocol as an international standard-setting body under the SPS Agreement. This step would formally introduce the rules of the Cartagena Protocol into the world trade system, alongside the standards of the Codex Alimentarius Commission, the International Plant Protection Convention and the International Office of Epizootics (Rivera-Torres 2003: 312–313). Beyond recognition of the current rules of the Cartagena Protocol, the proposal would shift the jurisdictional balance towards the Protocol by implicitly also recognizing its future decisions – which does not improve the prospects of the proposal being accepted. Second, the larger problem of the tension between the world trade system and multilateral environmental agreements with trade-restricting effects is part of the agenda of the Doha Round of trade negotiations. Any decision on the matter would immediately affect the interaction between the WTO and the Cartagena Protocol. However, both options for political decision-making are unlikely to be adopted because they face the well-known resistance of the United States, which is not a party to the CBD and its Cartagena Protocol (and is unlikely to join them in the foreseeable future – Brunnée 2004: 623–624). Progress on the larger issue of the relationship between the WTO and multilateral environmental agreements also faces scepticism and opposition from developing countries fearing a new wave of protection (e.g. Araya 2001; Gnath 2004).

In every case of political or judicial decision-making, the WTO will be the target of interaction through commitment originating from the Cartagena Protocol. Decision-making within the WTO will always occur in light of, and thus be affected by, the rules of the Protocol (and other international institutions such as the Codex Alimentarius Commission). Political decision-making requiring the broad support of WTO members can hardly be expected to produce decisions that are in open conflict with the Protocol, because the overwhelming majority of states are members of both institutions and this would drive them into incompatible commitments. Even non-members of the Cartagena Protocol, such as the United States, are likely to have little interest in making the two agreements incompatible, since this might harm the legitimacy of the WTO. Likewise, the dispute settlement mechanism is mandated to take into account general rules of international law and acts on behalf, and in the interest, of states parties. Because open conflict would endanger the legitimacy and governance capacity of the WTO, it should have an interest in working, to the extent possible, towards a *modus vivendi* rather than a possible escalation of tension (see Eggers and Mackenzie 2000: 540–542; Howse and Meltzer 2002). Indeed, the WTO dispute settlement organs have already developed a record of taking into account the provisions of multilateral environmental agreements and, to this end, they could exchange information with or ask for advisory opinions of the Cartagena Protocol, in particular its new Compliance Committee (see Eggers and Mackenzie 2000: 541–542; Howse and Meltzer 2002; Boisson de Chazournes and Mbengue 2004: 297–301).

Relevant judicial or political decision-making will affect the behavioural interaction between the WTO and the Cartagena Protocol. A significant redefinition of the delimitation of jurisdictions will inevitably influence domestic implementation (unless contracting parties choose to ignore and infringe valid rules). If the dispute settlement mechanism rejects bold restrictions on trade in GMOs/LMOs, WTO members will have to adapt their measures accordingly or else face trade sanctions. On the other hand, if the dispute settlement mechanism accepts a broad range of domestic measures to ensure biosafety, this may encourage originally hesitating countries to strengthen their restrictions on GMO imports. A political recognition of Cartagena rules within the world trade system can be expected to have a similar effect. In any event, a further specification of the delimitation of jurisdictions within the WTO would have to be taken into account under, and would thus influence, the Cartagena Protocol.

4.3. Possible further action under the Cartagena Protocol

Within the Cartagena Protocol, a further delimitation of jurisdictions under the WTO might be responded to by its own judicial or political decision-making. A judicial decision could result from a legal challenge of a country's regulation of trade in GMOs/LMOs under the compliance mechanism of the Cartagena Protocol established in 2004 (CBD 2004; Mackenzie 2004: 272–273). Triggering this compliance procedure constitutes a strategic policy option as a counter-measure by biosafety interests against a WTO challenge, especially because a party may trigger the procedure against itself. If its own domestic measures are challenged under WTO law, a party interested in biosafety could thereby seek to obtain the support of the biosafety regime on the matter and thereby increase the pressure on the WTO dispute settlement bodies to take into account and respect the Cartagena Protocol. With both (quasi-)judicial processes investigating the same case, pressure would rise to enter into some (informal) inter-judicial exchange to ensure compatible findings. There is little reason for supporters of either the Cartagena Protocol or the WTO to trigger the Cartagena compliance procedure *in order to strengthen their stance vis-à-vis the other side*. Relevant political decisions in response to the WTO could be taken by the Conference of the Parties to the Cartagena Protocol. Interested parties could seek adoption of additional rules to further define the jurisdictional delimitation of the regime vis-à-vis the WTO, for example with respect to risk assessment and precaution, as appropriate.

The scope for such judicial and political decision-making will be limited in turn by interaction through commitment originating from the WTO. It is hardly conceivable that the Conference of the Parties to the Cartagena Protocol or its Compliance Committee would ignore the WTO commitments of the vast majority of parties. Moreover, the preamble of the Protocol virtually instructs those interpreting the agreement – be it the Conference of the Parties or the Compliance Committee – to take into account the WTO regulations (section 3.2), which would include any decisions taken by the WTO on relevant matters. In these circumstances, initiatives under the Cartagena Protocol can be expected not to move the jurisdictional boundaries significantly in favour of biosafety but primarily to counter attempts by the world trade system to privilege trade interests (or carefully further to specify the existing balance).

As in the case of the WTO, relevant judicial or political decision-making can be expected to affect the behavioural interaction between the WTO and the Cartagena Protocol. First of all, contracting parties must adjust their domestic implementation measures to the new rules. Measures ruled out under the Protocol can no longer be sincerely applied and will almost certainly be considered as a violation of WTO commitments. And measures explicitly accepted under the Protocol will tend to be adopted even by parties that were originally hesitant because they feared trade conflicts.

Also in line with the effects of further decision-making on the matter within the WTO, secondary rules adopted under the Cartagena Protocol would modify the decision situation within the WTO. These rules will have to be taken into account in subsequent judicial and political decision-making and thus create a further case of interaction through commitment of the jurisdictional delimitation type. This could be followed by further decision-making within the WTO that would similarly influence the conditions for further decision-making under the Cartagena Protocol, and so on. The potential for the continuation of this feedback process is constrained only by the limited room that is available in the two regimes for advancing decision-making while staying compatible with the commitments of the other side.

5. Conclusion

The separate exploration of several relevant cases of interaction reveals the particularities of the complex interdependence of the WTO and the Cartagena Protocol. Our conceptual approach of disaggregating complex interaction situations into individual cases of interaction allows for the clear identification of causal relationships between the two institutions. The varying influence exerted by the WTO on the biosafety regime, and vice versa, becomes clearly visible. The analysis of the causal mechanism driving each case of interaction allows for the identification of crucial factors that have shaped the interaction as well as its consequences. By re-aggregating the individual cases, we acquire a broader picture of the driving forces and consequences at work. Since many of the factors underlying and shaping the interaction are likely to remain influential, their identification provides a solid basis for assessing the future relationship of the two regimes and the policy options available in this respect.

The overall interaction appears as a stepwise delimitation of jurisdictional authority. It is composed of several separate interaction cases that follow the causal mechanism of interaction through commitment and, more specifically, concern jurisdictional delimitation issues. On the one side, the WTO agreements of 1994 restricted the ability of the Cartagena Protocol to regulate trade in GMOs/LMOs. On the other side, the Cartagena Protocol has limited the scope for further regulation of this area by the WTO. The stepwise jurisdictional delimitation has been driven by the institutions' different objectives, as supported by two groups of states and policy communities as well as by the interest of states parties to both institutions in avoiding incompatible commitments.

Eventually, the relationship between the two agreements will be worked out further in their implementation "on the ground". Although future rule-making within either of the two institutions will have to take into account the commitments entered into by the member states of the other institution, the momentum of the future interdependence between the two institutions will largely be determined by behavioural interaction. Because domestic implementation of one institution will simultaneously affect the performance of the other institution, countries regulating trade in GMOs/LMOs will, within the existing margin for interpretation, decide on the exact balance between the objectives of free trade and biosafety. With the Cartagena Protocol having addressed the relationship with the WTO, its domestic implementation, which is only just beginning, is likely to be the focal point of future interaction with the world trade system.

The delimitation of jurisdictions of the WTO and the Cartagena Protocol has developed in the absence of centralized coordination. Since there is no suitable overarching international institution that could accommodate conflicting commitments, the members of multilateral treaty systems resort to collective decision-making within either of the institutions involved, while taking into account the objectives of the other institution. The transmission belt introducing external objectives into the internal decision processes of either institution is the joint membership of both institutions of the vast majority of parties. Twin members have an interest in both the regulatory objectives, and will therefore tend to avoid incompatible commitments. Hence, in spite of the tension between the objectives of the WTO and the Cartagena Protocol and despite the diverging interests of the member states as to the appropriate balance between them, certain features of the system of international governance drive the institutions towards accommodation even in the absence of a coordinating institution. As a result, there is a good chance that the two regimes will develop further in consistent ways in the future.

However, the largely successful delimitation of jurisdictions does not imply that the balance found is necessarily to the liking of all the actors involved. Jurisdictions can be delimited in different ways with different effects on outcomes. The exact balance struck is largely a matter of the distribution of power between the institutions involved. The trade side successfully secured a first-mover advantage by structuring the regulatory field through the WTO agreements of 1994, most importantly the SPS Agreement, so that biosafety negotiators had to operate from the very beginning in this "shadow of the WTO". Yet, the seemingly weak Cartagena Protocol showed surprising strength in assuming regulatory authority from the allegedly powerful WTO and in exploiting the remaining room for manoeuvre. As a result, effective protection of biological safety has gained support. Whether the resulting balance is sufficient for this purpose will be seen only through the implementation of the two agreements.

Notes

1. For a number of concrete case studies, see Rosendal (2000, 2001); Oberthür (2001); Andersen (2002); Jacquemont and Caparrós (2002); Oberthür and Gehring (2003).
2. See e.g. contributions in Adler (2000); Eggers and Mackenzie (2000); Bail et al. (2002); Burgiel (2002); Safrin (2002b); Rivera-Torres (2003).
3. The causal mechanisms are discussed in more detail in Oberthür and Gehring (2006b).
4. A more detailed discussion of the sub-types of the causal mechanisms and their characteristics can be found in Gehring and Oberthür (2006b).
5. For relevant analyses of the WTO agreements, see Eggers and Mackenzie (2000); Howse and Meltzer (2002); Safrin (2002b); Rivera-Torres (2003).
6. See Raustiala and Victor (2004) for a similar argument regarding the "regime complex for plant genetic resources".
7. Both institutions may actually be seen as belonging to a larger "regime complex" (Raustiala and Victor 2004), including, *inter alia*, the International Plant Protection Convention and the Codex Alimentarius Commission. For the importance of these two institutions for the conflict over GMO trade, see for example Homeyer (2006). Our focus here is on the two principal international institutions involved in this conflict – the WTO and the biosafety regime.

REFERENCES

Adler, Jonathan H. (2000), "The Cartagena Protocol and Biological Diversity: Biosafe or Bio-Sorry", *Georgetown International Environmental Law Review* 12(3): 761–777.

Alfonso, Margarida (2002), "The Relationship with Other International Agreements: An EU Perspective", in Christoph Bail, Robert Falkner and Helen Marquard (eds), *The Cartagena Protocol on Biosafety: Reconciling Trade in Biotechnology with Environment and Development?* London: RIIA/Earthscan, pp. 423–437.

Andersen, Regine (2002), "The Time Dimension in International Regime Interplay", *Global Environmental Politics* 2(3): 98–117.

Araya, Mónica (2001), "Environmental Dilemmas on the Road to Doha: Winning Southern Support for Greening the WTO", in Liane Schalatek (ed.), *Trade and Environment, the WTO, and Multilateral Environmental Agreements*. Washington, DC: Woodrow Wilson Center for Scholars, Heinreich Böll Foundation & the National Wildlife Federation, pp. 109–118.

Archer, Margaret S. (1985), "Structuration versus Morphogenesis", in S. N. Eisenstadt and H. J. Helle (eds), *Vol. 1 of Macro-Sociological Theory. Perspectives on Sociological Theory*. London: Sage, pp. 58–88.
Bail, Christoph, Robert Falkner and Helen Marquard, eds (2002), *The Cartagena Protocol on Biosafety: Reconciling Trade in Biotechnology with Environment and Development?* London: RIIA/Earthscan.
Boisson de Chazournes, Laurence and Makane Moïse Mbengue (2004), "GMOs and Trade: Issues at Stake in the EC Biotech Dispute", *Review of European Community and International Environmental Law* 13(3): 289–305.
Brack, Duncan (2002), "Environmental Treaties and Trade: Multilateral Environmental Agreements and the Multilateral Trading System", in Gary P. Sampson and W. Bradnee Chambers (eds), *Trade, Environment, and the Millennium*. Tokyo: United Nations University Press, pp. 321–352.
Breitmeier, Helmut (2000), "Complex Effectiveness, Regime Externalities and Interaction (Working Group III)", in Jorgen Wettestad (ed.), *Proceedings of the 1999 Oslo Workshop of the Concerted Action Network on the Effectiveness of International Environmental Regimes*. Oslo: Fridtjof Nansen Institute, pp. 45–48.
Brunnée, Jutta (2004), "The United States and International Environmental Law: Living with an Elephant", *European Journal of International Law* 15(4): 617–649.
Burgiel, Stanley W. (2002), "The Cartagena Protocol on Biosafety: Taking the Steps from Negotiation to Implementation", *Review of European Community and International Environmental Law* 11(1): 53–61.
Carlsnæs, Walter (1992), "The Agent–Structure Problem in Foreign Policy Analysis", *International Studies Quarterly* 36(3): 245–270.
Cartagena Protocol (2000), *Cartagena Protocol on Biosafety to the Convention on Biological Diversity: Text and Annexes*. Montreal: Secretariat of the Convention on Biological Diversity; available at ⟨http://www.cbd.int/doc/legal/cartagena-protocol-en.pdf⟩ (accessed 5 July 2007).
CBD [Convention on Biological Diversity] (2004), "Procedures and Mechanisms on Compliance under the Cartagena Protocol on Biosafety", in *Report of the First Meeting of the Conference of the Parties Serving as the Meeting of the Parties to the Protocol on Biosafety*. UN Doc. UNEP/CBD/BS/COP-MOP/1/15, 14 April, pp. 65–68.
Checkel, Jeffrey T. (1998), "The Constructivist Turn in International Relations Theory", *World Politics* 50(3): 324–348.
Clapp, Jennifer (1994), "Africa, NGOs, and the International Toxic Waste Trade", *Journal of Environment and Development* 3(2): 17–46.
Coleman, James (1990), *Foundations of Social Theory*. Cambridge, MA: Belknap Press.
Cottier, Thomas (2002), "Implications for Trade Law and Policy: Towards Convergence and Integration", in Christoph Bail, Robert Falkner and Helen Marquard (eds), *The Cartagena Protocol on Biosafety: Reconciling Trade in Biotechnology with Environment and Development?* London: RIIA/Earthscan, pp. 467–481.
Eggers, Barbara and Ruth Mackenzie (2000), "The Cartagena Protocol on Biosafety", *Journal of International Economic Law* 3(3): 525–543.

Elster, Jon (1989), *Nuts and Bolts for the Social Sciences*. Cambridge: Cambridge University Press.

Falkner, Robert (2000), "Regulating Biotech Trade: The Cartagena Protocol on Biosafety", *International Affairs* 76(2): 299–313.

Falkner, Robert (2002), "Negotiating the Biosafety Protocol. The International Process", in Christoph Bail, Robert Falkner and Helen Marquard (eds), *The Cartagena Protocol on Biosafety: Reconciling Trade in Biotechnology with Environment and Development?* London: RIIA/Earthscan, pp. 3–22.

Gehring, Thomas and Sebastian Oberthür (2004), "Exploring Regime Interaction: A Framework for Analysis", in Arild Underdal and Oran R. Young (eds), *Regime Consequences: Methodological Challenges and Research Strategies*. Dordrecht: Kluwer, pp. 247–269.

Gehring, Thomas and Sebastian Oberthür (2006), "Comparative Empirical Analysis and Ideal Types of Institutional Interaction", in Sebastian Oberthür and Thomas Gehring (eds), *Institutional Interaction in Global Environmental Governance: Synergy and Conflict among International and EU Policies*. Cambridge, MA: MIT Press.

Gnath, Katharina (2004), *The WTO and Environment: ... More Words Than Deeds? On the Relationship between the WTO and the Environment: Overview, Latest Developments and Assessment*. Berlin: Deutsche Gesellschaft für Auswärtige Politik; available at ⟨http://www.weltpolitik.net/Sachgebiete/Weltwirtschaft%20und%20Globalisierung/Institutionen%20und%20Akteure/WTO/Analysen/wto_env.html⟩ (accessed 10 July 2007).

Graff, Laurence (2002), "The Precautionary Principle", in Christoph Bail, Robert Falkner and Helen Marquard (eds), *The Cartagena Protocol on Biosafety: Reconciling Trade in Biotechnology with Environment and Development?* London: RIIA/Earthscan, pp. 410–422.

Gupta, Aarti (2001), "Advance Informed Agreement: A Shared Basis for Governing Trade in Genetically Modified Organisms?", *Indiana Journal of Global Legal Studies* 9(1): 265–281.

Haas, Peter M., Robert O. Keohane and Marc A. Levy, eds (1993), *Institutions for the Earth. Sources of Effective International Environmental Protection*. Cambridge, MA: MIT Press.

Hedström, Peter and Richard Swedberg (1998), "Social Mechanisms. An Introductory Essay", in Peter Hedström and Richard Swedberg (eds), *Social Mechanisms: An Analytical Approach to Social Theory*. Cambridge: Cambridge University Press, pp. 1–31.

Héritier, Adrienne (1996), "The Accommodation of Diversity in the European Policy-making Process and Its Outcomes. Regulatory Policy as a Patchwork", *Journal of European Public Policy* 3(2): 149–167.

Homeyer, Ingmar von (2006), "The EU Deliberate Release Directive: Environmental Precaution versus Trade and Product Regulation", in Sebastian Oberthür and Thomas Gehring (eds), *Institutional Interaction in Global Environmental Governance: Synergy and Conflict among International and EU Policies*. Cambridge, MA: MIT Press.

Howse, Robert and Joshua Meltzer (2002), "The Significance of the Protocol for WTO Dispute Settlement", in Christoph Bail, Robert Falkner and Helen

Marquard (eds), *The Cartagena Protocol on Biosafety: Reconciling Trade in Biotechnology with Environment and Development?* London: RIIA/Earthscan, pp. 482–496.

Jacquemont, Frédéric and Alejandro Caparrós (2002), "The Convention on Biological Diversity and the Climate Change Convention 10 Years after Rio: Towards a Synergy of the Two Regimes?", *Review of European Community and International Environmental Law* 11(2): 139–180.

Keohane, Robert O. (1984), *After Hegemony. Cooperation and Discord in the World Political Economy.* Princeton, NJ: Princeton University Press.

Lanchbery, John (2006), "The Convention on International Trade in Endangered Species of Wild Fauna and Flora (CITES): Responding to Calls for Action from Other Nature Conservation Regimes", in Sebastian Oberthür and Thomas Gehring (eds), *Institutional Interaction in Global Environmental Governance: Synergy and Conflict among International and EU Policies.* Cambridge, MA: MIT Press.

Levy, Marc A., Oran R. Young and Michael Zürn (1995), "The Study of International Regimes", *European Journal of International Relations* 1(3): 267–330.

Mackenzie, Ruth (2004), "The Cartagena Protocol after the First Meeting of the Parties", *Review of European Community and International Environmental Law* 13(3): 270–278.

Martin, Lisa L. (1993), "The Rational State Choice of Multilateralism", in John Ruggie (ed.), *Multilateralism Matters. The Theory and Praxis of an Institutional Form.* New York: Columbia University Press, pp. 91–121.

Mattli, Walter (2003), "Setting International Standards: Technological Rationality or Primacy of Power", *World Politics* 56(1): 1–42.

Meinke, Britta (2002), *Multi-Regime-Regulierung. Wechselwirkungen zwischen globalen und regionalen Umweltregimen.* Darmstadt: Deutscher Universitäts-Verlag.

Miles, Edward L., Arild Underdal, Steinar Andresen, Jørgen Wettestad, Jon Birger Skjærseth and Elaine M. Carlin (2002), *Environmental Regime Effectiveness: Confronting Theory with Evidence.* Cambridge, MA: MIT Press.

Oberthür, Sebastian (2001), "Linkages between the Montreal and Kyoto Protocols: Enhancing Synergies between Protecting the Ozone Layer and the Global Climate", *International Environmental Agreements: Politics, Law and Economics* 1(3): 357–377.

Oberthür, Sebastian and Thomas Gehring (2003), "Investigating Institutional Interaction: Towards a Systematic Analysis", paper presented at the International Studies Association Convention, Portland, Oregon, 12 February to 1 March.

Oberthür, Sebastian and Thomas Gehring, eds (2006a), *Institutional Interaction in Global Environmental Governance: Synergy and Conflict among International and EU Policies.* Cambridge, MA: MIT Press.

Oberthür, Sebastian and Thomas Gehring (2006b), "Conceptual Foundations of Institutional Interaction", in Sebastian Oberthür and Thomas Gehring (eds), *Institutional Interaction in Global Environmental Governance: Synergy and Conflict among International and EU Policies.* Cambridge, MA: MIT Press.

Oberthür, Sebastian and Hermann E. Ott, in collaboration with Richard G. Tarasofsky (1999), *The Kyoto Protocol. International Climate Policy for the 21st Century*. Berlin: Springer.

Palmer, Alice, Beatrice Chaytor and Jacob Werksman (2006), "Interactions between the World Trade Organisation and International Environmental Regimes", in Sebastian Oberthür and Thomas Gehring (eds), *Institutional Interaction in Global Environmental Governance: Synergy and Conflict among International and EU Policies*. Cambridge, MA: MIT Press.

Pythoud, Francois and Urs P. Thomas (2002), "The Cartagena Protocol on Biosafety", in Philippe G. Le Prestre (ed.), *Governing Global Biodiversity: The Evolution and Implementation of the Convention on Biological Diversity*. Aldershot: Ashgate, pp. 39–56.

Raustiala, Kal and David G. Victor (2004), "The Regime Complex for Plant Genetic Resources", *International Organization* 58(2): 277–309.

Risse, Thomas (2000), "Let's Argue! Communicative Action in World Politics", *International Organization* 54(1): 1–39.

Risse-Kappen, Thomas (1994), "Ideas Do Not Float Freely. Transnational Coalitions, Domestic Structures, and the End of the Cold War", *International Organization* 48(2): 185–214.

Rivera-Torres, Olivette (2003), "The Biosafety Protocol and the WTO", *Boston College International and Comparative Law Review* 26(2): 263–323.

Rosendal, G. Kristin (2000), *The Convention on Biological Diversity and Developing Countries*. Dordrecht: Kluwer Academic.

Rosendal, Kristin (2001), "Impacts of Overlapping International Regimes: The Case of Biodiversity", *Global Governance* 7(1): 95–117.

Safrin, Sabrina (2002a), "The Relationship with Other Agreements: Much Ado about a Savings Clause", in Christoph Bail, Robert Falkner and Helen Marquard (eds), *The Cartagena Protocol on Biosafety: Reconciling Trade in Biotechnology with Environment and Development?* London: RIIA/Earthscan, pp. 438–454.

Safrin, Sabrina (2002b), "Treaties in Collision? The Biosafety Protocol and the World Trade Organization Agreements", *American Journal of International Law* 96(3): 606–627.

Schelling, Thomas (1998), "Social Mechanisms and Social Dynamics", in Peter Hedström and Richard Swedberg (eds), *Social Mechanisms: An Analytical Approach to Social Theory*. Cambridge: Cambridge University Press, pp. 32–44.

Schoenbaum, Thomas (2002), "International Trade and Environmental Protection", in Patricia Birnie and Alan Boyle (eds), *International Law and the Environment*, 2nd edn. Oxford: Oxford University Press, pp. 697–750.

Selin, Henrik and Stacy D. VanDeveer (2003), "Mapping Institutional Linkages in European Air Pollution Politics", *Global Environmental Politics* 3(3): 14–46.

Shapiro, Martin (1981), *Courts: A Comparative and Political Analysis*. Chicago: University of Chicago Press.

Shaw, Sabina and Risa Schwartz (2002), "Trade and Environment in the WTO: State of Play", *Journal of World Trade* 36(1): 129–154.

Simon, Herbert A. (1972), "Theories of Bounded Rationality", in Charles B. McGuire and Roy Radner (eds), *Decision and Organization*. Amsterdam: North-Holland, pp. 161–176.

Skjærseth, Jon Birger (2006), "Protecting the North-East Atlantic: One Problem, Three Institutions", in Sebastian Oberthür and Thomas Gehring (eds), *Institutional Interaction in Global Environmental Governance: Synergy and Conflict among International and EU Policies*. Cambridge, MA: MIT Press.

SPS Agreement (1994), *Agreement on the Application of Sanitary and Phytosanitary Measures. Annex 1A to the Final Act Embodying the Results of the Uruguay Round of Multilateral Trade Negotiations*, Marrakesh, 15 April 1994; available at ⟨http://www.wto.org/English/docs_e/legal_e/15-sps.pdf⟩ (accessed 5 July 2007).

Stein, Arthur (1982), "Coordination and Collaboration. Regimes in an Anarchic World", *International Organization* 36(2): 299–324.

Stokke, Olav Schram (2001a), "Conclusions", in Olav Schram Stokke (ed.), *Governing High Seas Fisheries: The Interplay of Global and Regional Regimes*. Oxford: Oxford University Press, pp. 329–360.

Stokke, Olav Schram (2001b), *The Interplay of International Regimes: Putting Effectiveness Theory to Work*, FNI Report 14/2001. Lysaker: Fridtjof Nansen Institute.

Stoll, Peter-Tobias (2000), "Controlling the Risks of Genetically Modified Organisms: The Cartagena Protocol on Biosafety and the SPS Agreement", *Yearbook of International Environmental Law 1999* 10: 82–119.

TBT Agreement (1994), *Agreement on Technical Barriers to Trade. Annex to the Final Act Embodying the Results of the Uruguay Round of Multilateral Trade Negotiations*, Marrakesh, 15 April 1994; text available at ⟨http://www.wto.org/english/docs_e/legal_e/17-tbt.pdf⟩ (accessed 5 July 2007).

Underdal, Arild (2004), "Methodological Challenges in the Study of Regime Effectiveness", in Arild Underdal and Oran R. Young (eds), *Regime Consequences: Methodological Challenges and Research Strategies*. Dordrecht: Kluwer, pp. 27–48.

Werksman, Jacob (2005), "The Negotiation of a Kyoto Compliance System", in Olav Schram Stokke, Jon Hovi and Geir Ulfstein (eds), *Implementing the Climate Regime: International Compliance*. London: Earthscan, pp. 17–38.

Yee, Albert S. (1996), "The Causal Effects of Ideas on Politics", *International Organization* 50(1): 69–108.

Young, Oran R. (1992), "The Effectiveness of International Institutions: Hard Cases and Critical Variables", in James N. Rosenau and Ernst-Otto Czempiel (eds), *Governance without Government: Order and Change in World Politics*. Cambridge: Cambridge University Press, pp. 160–194.

Young, Oran R. (1996), "Institutional Linkages in International Society: Polar Perspectives", *Global Governance* 2(1): 1–24.

Young, Oran R., ed. (1999), *The Effectiveness of International Environmental Regimes: Causal Connections and Behavioral Mechanisms*. Cambridge, MA: MIT Press.

Young, Oran R. (2002), *The Institutional Dimensions of Environmental Change: Fit, Interplay, and Scale*. Cambridge, MA: MIT Press.

Young, Oran R., with contributions from Arun Agrawal, Leslie A. King, Peter H. Sand, Arild Underdal and Merrilyn Wasson (1999), *Science Plan: Institutional Dimensions of Global Environmental Change*, IHDP Report No. 9. Bonn: International Human Dimensions Programme on Global Environmental Change.

Zedan, Hamdallah (2002), "The Road to the Biosafety Protocol", in Christoph Bail, Robert Falkner and Helen Marquard (eds), *The Cartagena Protocol on Biosafety: Reconciling Trade in Biotechnology with Environment and Development?* London: RIIA/Earthscan, pp. 23–33.

Part III
Conclusion

6

Deriving insights from the case of the WTO and the Cartagena Protocol

Oran R. Young

1. Introduction

There is nothing surprising about the growth of interplay between and among regimes or issue-specific governance systems operating at the international level. Recent decades have witnessed a rapid rise in the number of such international governance systems; they now number in the hundreds and address a broad spectrum of issues ranging from environmental and economic concerns to matters of public health, human rights and international security. One response to this striking development centres on efforts to understand the life histories – including issues of formation, implementation and evolution – of individual regimes. But, as a number of well-informed observers have noted, we can confidently predict that many of these institutional arrangements will interact with one another, giving rise to relationships that may have important consequences for the capacity of individual governance systems to attain their goals and to play key roles in coming to terms with the problems that led to their creation. Growth in the need to understand the nature and consequences of institutional interplay is therefore unavoidable.

Difficulties in pinpointing the core elements and locating the boundaries of individual regimes can complicate efforts to analyse institutional interplay in specific cases. But there is no reason for such complications at the margins to deter us from identifying a sizeable universe of cases in which the occurrence of institutional interplay is manifest. There are, in other words, numerous cases in which two or more regimes that have

Institutional interplay: Biosafety and trade, Young, Chambers, Kim and ten Have (eds), United Nations University Press, 2008, ISBN 978-92-808-1148-3

separate histories, are codified in different international agreements and give rise to distinct implementation practices interact with one another with consequences that cannot be ignored. The case we have chosen to examine from a variety of perspectives in this book – interactions between the Cartagena Protocol on Biosafety and the World Trade Organization (WTO) – exemplifies the class of situations of interest to those who study institutional interplay. In some respects, this is a relatively complex case; both sides of this interaction involve what we can call compound regimes or, in other words, arrangements encompassing two or more components. In the case of the WTO, for instance, we need to consider the Agreement on the Application of Sanitary and Phytosanitary Measures (SPS Agreement), the Agreement on Trade-Related Intellectual Property Rights (TRIPs Agreement), the Agreement on Technical Barriers to Trade (TBT Agreement) and the Codex Alimentarius Commission,[1] as well as the core provisions of the General Agreement on Tariffs and Trade (GATT) itself. For its part, the Cartagena Protocol is embedded in a larger system that features the Convention on Biological Diversity and addresses the overarching problem of maintaining biological diversity in general terms. As Oberthür and Gehring in particular make clear (Chapter 5 in this volume), this complexity poses methodological problems for those seeking to analyse institutional interplay. At its core, however, this case clearly illustrates many features of interplay; it provides a good vehicle for an effort to shed light on institutional interplay by moving back and forth between theory and practice.

The strategy we employed in putting together this volume was straightforward. We asked a number of scholars to look at the same case – the WTO–Protocol interaction – from the perspective of different approaches to the analysis of interplay as a means of drawing attention to similarities and differences among these approaches. Throughout this exercise, we have focused attention on a few critical questions pertaining to institutional interplay. What is the nature of institutional interplay and how does it arise? What are the consequences of institutional interplay? And what, if anything, should we do about interplay in cases where it is found to have important consequences for the performance of one or more of the issue-specific regimes involved? Our hope is that efforts to understand a single important case from a variety of analytic angles will prove intellectually constructive in the sense that it gives rise to insights about interplay that are both productive in advancing the scientific study of environmental governance and helpful to those responsible for administering individual regimes and for making policy decisions about the (re)formation of governance systems operating in a variety of issue areas.

2. Limits to taxonomy

So far, efforts to illuminate the nature of institutional interplay have focused on concept formation and have produced results that are largely taxonomic in character. This has triggered a proliferation of conceptual distinctions directing attention to aspects of interplay that individual analysts have found interesting. Thus, researchers speak of forms of interplay (e.g. horizontal/vertical or symmetrical/asymmetrical), drivers of interplay (e.g. power, interests, knowledge), mechanisms of interplay (e.g. utilitarian, normative, ideational), targets of interplay (e.g. norms, rules), results of interplay (e.g. outputs, outcomes, impacts) and consequences of interplay (e.g. synergy, conflict).[2] And there is little to stop this taxonomic proliferation from continuing during the foreseeable future. Those interested in institutional interactions can and surely will think of additional dimensions of interplay that seem helpful from one point of view or another.

There is nothing inherently wrong with efforts to develop concepts and conceptual frameworks to be used in addressing a newly emerging topic such as institutional interplay in international society. Some of the resultant distinctions will surely prove helpful. Most would agree, I expect, that it is useful to differentiate between horizontal interplay involving interactions between or among regimes operating at the same level of social organization and vertical interplay involving interactions between or among governance systems operating at different levels of social organization (Young et al. 1999). Whether implicit or explicit, the presumption here is that the dynamics of the two types of interplay differ in ways that make it impossible to develop a single set of propositions that apply equally well to both types of interplay. In a recent essay, I have argued that cross-level interactions among scale-dependent regimes or governance systems constitute a distinctive sub-set even within the domain of vertical interplay (Young 2006). Yet the horizontal/vertical distinction is not critical to understanding interplay between the WTO and the Protocol; this case is a relatively straightforward instance of horizontal interplay.

In my judgement, the benefits of taxonomic proliferation are marginal at best when it comes to the search for insights into the origins and consequences of interplay in a case such as the interactions between the Cartagena Protocol and the WTO. This is partly a consequence of taxonomic sloppiness. Many of the distinctions we have introduced so far do not yield categories that are mutually exclusive and exhaustive. This is certainly the case with regard to the distinction between functional and political interplay, a distinction that I must assume a fair share of the responsibility for introducing as a feature of the Science Plan for the project on the Institutional Dimensions of Global Environmental Change (Young et al. 1999). From a purely taxonomic perspective, then, our efforts to draw distinctions among various types of interplay leave much to be desired.

Even more important, however, is the lack of clear-cut links between many of our conceptual distinctions relating to institutional interplay and theoretical concerns that can help us to understand the origins and consequences of interplay or that can be clarified or refined as a result of research dealing with actual instances of interplay. By way of comparison, we distinguish between common-pool resources and private goods because we think different governance systems are needed to manage human uses of these goods in a sustainable manner (Ostrom et al. 2002). We distinguish between coordination problems and collaboration problems because we expect that they will differ fundamentally with regard to problems of compliance (Stein 1983). We distinguish between cap-and-trade arrangements and command-and-control regulations because we believe they will produce results that differ substantially in terms of common measures of efficiency (Tietenberg 2002). Of course, these expectations may turn out to be incorrect or, more likely, prove tenable under some conditions but not others. Thus, we have expended a good deal of time and energy on efforts to understand the nature of the links between common-pool resources and the occurrence of the tragedy of the commons. But no one doubts the importance of the conceptual distinctions underlying these expectations.

With regard to institutional interplay, the problem is that the links between many of our taxonomic distinctions and important theoretical concerns are, at best, unclear or underdeveloped. We do not have clearly stated expectations that reflect theoretically important concerns regarding the drivers, mechanisms or targets of institutional interplay. And we are unlikely to outgrow this problem if we continue to introduce conceptual distinctions that seem on the surface to offer insights but that turn out on more careful examination to be unrelated to important theoretical concerns regarding the role of institutions in causing or addressing major environmental problems. The appropriate remedy, in my judgement, is to adopt a strategy that curbs taxonomic proliferation and that subjects specific distinctions pertaining to the nature and consequences of institutional interplay to theoretically informed evaluation.

3. Theory-driven expectations

How should we proceed in the light of this conclusion? In this section and in the sections to follow, I argue that there is much to be gained by starting with two simple distinctions, demonstrating their theoretical importance, using them to illuminate the WTO and the Protocol case, and drawing on the evidence from the case to critique and refine the distinctions. The first of these distinctions separates cases of institutional interplay that are unintended from those that are the products of intentional actions on the part of important players. Unintended interplay centres on the occurrence of side effects that no one seeks to produce and that, more often than not, actors responsible for the creation and implementation of governance systems designed to solve specific problems do not foresee. Side effects of this sort may take a wide range of forms. Because chlorofluorocarbons (CFCs) are greenhouse gases, for example, efforts to reduce or eliminate the use of CFCs as a means of avoiding the depletion of stratospheric ozone are beneficial to efforts to avoid anthropogenic interference in the Earth's climate system. Because marine mammals feed on various species of fish, on the other hand, efforts to increase sustainable harvests in certain fisheries may interfere with or even undermine arrangements designed to protect marine mammals. As those who think about such things in terms of the idea of externalities will note, and as the preceding examples suggest, side effects can be either positive or negative with respect to their consequences for efforts to solve specific problems (Mishan 1982).

As the number of distinct regimes or governance systems operating in a given social space grows, it is predictable that both the frequency and the scope of these unintended side effects will increase. In fact, some simple arithmetic suggests that – other things being equal – institutional interplay of this sort will grow at an exponential pace. If the resultant problems were easy to fix and the occasional benefits easy to amplify, we might be justified in paying little attention to the growth of these institutional side effects. But this is not generally the case. As those who labour to internalize negative externalities know all too well, addressing such issues effectively is apt to be difficult and for several different reasons. Calculating the relevant costs or benefits is almost always controversial. Winners are often few in number and well organized, whereas large numbers of losers find it difficult to organize. The political system is apt to have a hard time addressing issues of this sort whose distributive implications are unavoidable. Although those striving to build regimes to solve specific problems generally do not intend their actions to impinge on the efforts of others to solve their own problems, therefore, coming to terms with institutional interplay that takes the form of unintended side effects presents a complex challenge, one that is destined to grow with the increase in the number of discrete institutions operating in international society.

When institutional interplay is intentional in character, by contrast, we are confronted with a different set of concerns. Like unintended interplay, intended interplay can take a variety of forms. There are cases (e.g. monitoring various forms of air pollution or compliance with regulations designed to control such pollution) in which participants can benefit from economies of scale by linking separate regimes; the marginal cost of adding another pollutant to an existing monitoring effort will often be a fraction of the cost of establishing a separate system to monitor that pollutant. On the other hand, actors often seek to create new regimes to promote causes that seem unlikely to fare well in existing institutional settings. As I shall argue in the next section, this is an important feature of the Protocol and WTO story.

As in the case of institutional externalities, then, intended interplay may have positive or negative consequences with respect to efforts to solve specific problems. A number of analysts have used the term "synergy" to describe positive interactions, while speaking of "conflict" or "interference" in the case of negative interactions (Oberthür and Gehring 2006). But the challenges that arise in efforts to come to terms with intended – in contrast to unintended – interplay are distinct. Where synergy is possible, it should be a fairly straightforward matter to reap the benefits. There may be a need, in specific cases, to explore the options and to reach agreement on the allocation of benefits arising from cooperation. But the win–win character of such interactions provides a strong basis for expecting the relevant parties to find ways to exploit opportunities for achieving mutual gains. The truly difficult cases are those in which the parties endeavour to exploit institutional interplay to pursue their own ends or to thwart the plans of others. Those who push for the creation and implementation of regimes calling for ecosystem-based management of marine areas, for instance, can be expected to emerge as opponents of those who operate within the framework of regimes designed to produce maximum sustainable yields from species of interest to particular groups of harvesters. In many cases, the resultant conflict will become acute, since gains for supporters of one type of governance system will be interpreted as losses for those who espouse the other type of regime. In extreme cases, polarization of this sort can give rise to pure or zero-sum conflict.

Other things being equal, then, we should expect intended interplay that is conflictual to be the most difficult type of interplay to manage effectively. In such cases, the relevant players are likely to expend their time and energy devising ways to assert or enhance the dominance of their preferred approach to governance over the system(s) that others prefer. Intended interplay that is synergistic in character, by contrast, should be comparatively easy to manage effectively. Here the key issue is to facilitate collaboration that allows those involved to take advantage of mutually beneficial opportunities, while minimizing the transaction costs associated with the necessary coordination efforts. Although the prospect of reaping gains from trade does not guarantee success in this realm, it is reasonable to expect a high probability of success. On this account, unintended interplay should fall somewhere between these cases of synergy and conflict. Even when they involve negative externalities, such interactions are easier to come to terms with than cases of intended conflict. Yet the incentives to take advantage of positive externalities are not as easy to activate as those that arise in cases featuring intended synergy.

The distinction between unintended and intended interplay thus raises issues that are of theoretical importance and produces expectations that should help to understand the nature of the Protocol and WTO case. But there is a second, cross-cutting distinction that can help to explain both the character and the consequences of institutional interplay in the specific case of the WTO and the Protocol and, in the process, deepen our general understanding of institutional interplay. Whether it is unintended or intended, institutional interplay can be either shallow or deep. Shallow interplay involves interactions that are more or less superficial in nature; it is often possible to address them through the development of technical measures. Thus, some substitutes for CFCs are more climate friendly than others; the achievement of economies of scale in the operation of monitoring systems or compliance mechanisms is often a matter that experts can sort out without raising major policy issues; friction may arise from relatively technical concerns that do not threaten the success of either regime. Deep interplay, by contrast, goes far beyond such operational matters to address issues of principles, norms, discourses and, ultimately, values. Those who have developed the idea of embedded liberalism, for instance, take the view that two or more economic regimes – such as the international monetary and trade regimes – can operate in a synergistic manner in large part because they share a fundamental commitment to the goal of promoting a liberal, market-based world economy that thrives on the growth of trade and flows of foreign direct investment (Ruggie 1983). When governance systems clash at this level, however, it is apparent that the problems of managing the resultant interplay will be altogether different in nature. The interplay between regimes calling for conservation in the sense of sustainable harvesting of living resources and those based on the value of preservation or, in other words, avoiding intentional killing of wildlife exemplifies this case.[3]

The shallow/deep distinction is useful whether we are concerned with unintended or with intended interplay. As the density of institutions increases, the incidence of inadvertent inconsistencies between or among specific procedures or regulations promulgated to operationalize individual regimes will grow. There is simply no way to do a thorough or exhaustive check for consistency in this realm once a large number of distinct governance systems are operative in the same social space (e.g. international society). For the most part, however, these inconsistencies will not pose serious problems. Administrative agencies or judicial bodies can make the necessary adjustments using standards or criteria devised especially to sort out unintended inconsistencies of this nature. Far more troubling are unintended tensions that arise from more fundamental sources. A prominent example that is pertinent to the case of the Protocol and the WTO arises from the very success of the trade regime. In essence, the goal of the trade regime is to promote economic growth and development by encouraging the free and unencumbered movement of goods and services across national boundaries. Yet economic growth as we know it today is a major source of environmental problems arising from over-exploitation of natural resources, habitat destruction and the dispersal of invasive species. It is not the intention of the trade regime to cause environmental harm; the problems at stake are simply unintended side effects of efforts on the part of well-intentioned actors to fulfil the goals of the trade regime. Finding ways to curtail these environmental side effects without impeding the growth of trade is a tall order. A large part of this problem is attributable to the fact that the interplay in question is deep rather than to the fact that it is unintended.

Parallel remarks are in order regarding the distinction between shallow and deep interactions in the context of intended interplay. Efforts to combine forces to benefit from economies of scale with regard to matters such as monitoring, compliance and dispute resolution may prove attractive to those responsible for operating two or more regimes (e.g. the separate regimes dealing with long-range transport of air pollution, ozone depletion and climate change). There is nothing trivial or unimportant about institutional interplay of this sort. Yet there is an important sense in which such interactions are shallow compared with interactions that arise from divergences involving underlying principles, norms and values. A growing number of cases involving interactions between trade regimes and regimes dedicated to the promotion of public health or the protection of the integrity of ecosystems exemplify deeper interactions that are intended in character. Because the cardinal principle of trade regimes calls for dismantling or minimizing impediments to trade, those who worry about the role of trade in the dispersal of invasive species or the spread of diseases are bound to have reservations about the consequences of trade regimes. Of course, it may be possible to address such side effects of free trade within the context of the trade regimes themselves. But the playing field is seldom level in such settings: the regimes in question are bound to privilege the case for expanding trade over the arguments of those who fear the impacts of various side effects of trade. Faced with this problem, advocates of values such as public health and biological diversity can and often will resort to the creation of separate regimes that become vehicles for defending and promoting their core values. The result, more often than not, is deep interplay that is intended in nature.

I will return to a consideration of outcomes likely to arise from such interactions in a later section. For now, the inference to be drawn from this discussion is that we can expect deep interplay – whether intended or not – to raise issues that are much more far-reaching than those associated with shallow interplay. This is not to say that deep interplay is always a bad thing. It is possible to imagine multiple regimes, such as the trade and monetary regimes, tied together by common norms and values in such a way that they reinforce each other at the level of day-to-day practice. Nonetheless, it is easy to see that the toughest problems arising from institutional interplay will emerge when adherents of different world views and value systems create regimes to promote their own ends and proceed to use these regimes intentionally to circumscribe the efforts of others or, at a minimum, to protect their own values from the inroads of supporters of other regimes. The result is institutional interplay that is deep, intentional and conflictual in nature. Without implying that other forms of interplay are uninteresting or unimportant, it follows that cases of interplay that fall into this category will become centres of attention in any social setting.

Figure 6.1 displays these relationships in graphical terms. The horizontal axis runs from unintended to intended interplay; the vertical axis runs from shallow to deep interplay. One immediate observation arising from

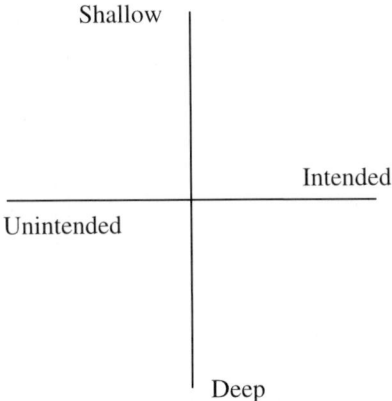

Figure 6.1 Dimensions of institutional interplay

this mode of presentation is that the unintended/intended and shallow/deep distinctions are not dichotomous in nature. There is room for some debate about such matters. Is interplay intended, for instance, in cases where such matters become a focus of debate in two-level games (Putnam 1988)? What about cases in which there is an element of intentionality in interactions but this is only a marginal or second- or third-order concern for most of the parties?

Nonetheless, the basic message of the figure is clear. As we move in a south-easterly direction in this two-dimensional space, the significance of interplay rises and the challenge of dealing with it in a constructive manner grows more severe. In the next section, I shall argue that the interplay between the Cartagena Protocol and the WTO lies deep in the lower right-hand corner of this figure (see also Schroeder, Chapter 3, and Sydnes, Chapter 4, in this volume). With this characterization in mind, I turn in section 5 to an analysis of the politics of this case of institutional interplay and to an assessment of the options available to those desiring to limit polarization over the introduction of genetically modified organisms and the disruptive impacts of such polarization on efforts to promote international cooperation more generally.

4. The case of the WTO and the Cartagena Protocol

The regime governing international trade is an integral component of a larger system of economic institutions created in the aftermath of the Great Depression and World War II and designed to promote the growth of a liberal world order (Irwin 2002). The rationale for a trade regime and the Bretton Woods institutions rests on several linked assumptions. Countries with open economies featuring competitive markets will engage in mutually beneficial trade that stimulates economic growth in the developed world and encourages economic development in the countries of the developing world. This in turn will give rise to an increasingly dense web of mutually beneficial relationships that will promote peace at the international level by making severe conflict – especially violent conflict – too costly and too risky for countries to embark upon in the absence of extreme provocation. The combination of economic prosperity and peace will contribute to the spread of democracy throughout international society.

The particular contribution of the World Trade Organization – and the General Agreement on Tariffs and Trade before it – to this strategy is to commit participants to the practice of free trade and to a determined and persistent effort to eliminate or at least to minimize barriers to trade through an ongoing series of multilateral negotiations, such as the Uruguay Round, which ended in 1994 with the creation of the WTO itself (Gilpin 2001). The emphasis throughout is on the liberalization of trade, not only through mutual agreements to reduce tariffs over time in a variety of sectors but also through a determined and persistent effort to identify and root out indirect or hidden arrangements motivated by political pressures within individual countries to cater to influential cconomic interests. Clearly embedded in the terms of the GATT from 1947 onward, this emphasis on eliminating indirect restraints on trade has become even more central to the trade regime under the terms of the WTO. Thus, the TBT Agreement focuses squarely on the desirability of phasing out domestic measures that do not rely on tariffs as such but that nevertheless prove effective in restricting or even prohibiting the growth of international trade. For its part, the SPS Agreement – incorporated into the trade regime as a product of the Uruguay Round – stresses the importance of avoiding any arbitrary or unjustified discrimination between countries, even when there are valid reasons to take actions necessary to protect "human, animal or plant life or health". In short, both the underlying principles and the ethos of this regime privilege trade liberalization and exert pressure on individual member countries to err on the side of minimizing restraints on trade in situations involving either uncertainty or unavoidable trade-offs.

Just as the trade regime reflects the preoccupation with building a prosperous world economy and a peaceful world order during the postwar era, the Cartagena Protocol on Biosafety – and the Convention on Biological Diversity in which it is embedded – reflects a set of concerns that moved onto centre stage during the 1990s and whose influence is currently growing. Broadly speaking, these concerns have to do with the expanding impacts of human actions on biophysical systems from the local level to the planetary level and the resultant awareness that we live now in a world of human-dominated ecosystems (Vitousek et al. 1997). More specifically, the emergence of biotechnology – not to be confused with the longstanding tradition of plant breeding – has given rise to a set of concerns about the (potential) impacts of genetically modified organisms (GMOs) on human health and on the resilience of the ecosystems in which humans live (Pollan 2001). There are legitimate concerns, for instance, about the impacts of the globalization of GMO crops on biological diversity, and there is reason to believe that a worldwide reliance on GMO crops could give rise to serious threats to human health. Some of these concerns are attributable more to uncertainty or a lack of knowledge (especially regarding long-term effects of GMO crops) than to certain knowledge about the nature and extent of harmful side effects associated with the rise of biotechnology. But, if anything, uncertainty coupled with understandable fears regarding impacts on human health has only magnified the concerns underlying the rights and rules articulated in the Protocol.

Exacerbating these concerns is a spreading sense among residents of the developed world and the developing world alike to the effect that the growth of international trade – and the rise of globalization more generally – is a development that works to the advantage of and is largely controlled by corporate interests around the world (Lang and Hines 1993). Perhaps the most dramatic manifestation to date of this questioning of the traditional doctrine regarding the benefits of free trade occurred in 1999 in the form of the anti-WTO riots in Seattle, which derailed efforts to launch a new round of trade negotiations under the auspices of the WTO and brought efforts to negotiate the terms of a new multilateral agreement on investments to a halt. But this was merely a surface expression of a much larger and deeper shift in opinion regarding the persuasiveness of the assumptions embedded in the complex of international economic institutions under contemporary conditions. It is beyond doubt that the negotiations that led to the signing of the Cartagena Protocol in January 2000 were influenced profoundly by this sea change in attitudes toward the international trade regime and the underlying premises in which it is rooted.

Needless to say, it is important to avoid exaggerating these differences in the fundamental premises on which the WTO and the Protocol rest. Those wishing to de-emphasize the importance of these differences – and many belong to this group – can point to a number of factors that can and sometimes do soften the clash between the principles, norms and discourses in which discussions of trade and biosafety are embedded (Gupta, Chapter 2 in this volume; Oberthür and Gehring, Chapter 5 in this volume). The provisions of the Protocol do not apply to GMO-based pharmaceuticals. Agricultural commodities intended for food, feed and processing are treated differently than GMO crops. Article XX of the GATT and similar provisions of the TBT and SPS agreements allow for exceptions to the general rules of the trade regime where these are "necessary to protect human, animal or plant life or health" or in situations "relating to the conservation of exhaustible natural resources" (Steinberg 2002). Taken at face value, these provisions may appear to offer sufficient latitude to respond sensitively and appropriately to the concerns of those who worry about the impact of GMO crops on human health and the resilience of ecosystems. But the exceptions are accompanied by injunctions about avoiding their use to promote arbitrary or unjustifiable discrimination or to disguise protectionist measures that are otherwise inadmissible. And, of course, the application of these provisions to specific situations takes place largely under the auspices of the WTO, a situation that is hardly conducive to allaying the concerns of those who worry about various aspects of biosafety.

Defenders of the WTO will argue with some justification that the trade regime is committed firmly to a policy of relying on the best available science to address issues relating to the environmental impacts of trade. And an examination of the practice of the SPS Agreement in using objective and accurate scientific data and of the role of the Codex Alimentarius Commission in providing high-quality scientific assessments of matters relating to food safety and animal and plant health provides some reassurance in these terms. However, the scientific community is in fact far from unanimous when it comes to formulating a proper approach to issues of this sort. One contingent stresses the role of experiments and the commitment to objectivity in the assessment of data; members of this group believe that science can produce objective knowledge providing a basis for making decisions in cases where various stakeholder groups call for restrictions on trade (Ruttan 1994; Andresen et al. 2000). Others both emphasize the difficulties in anticipating longer-term and systemic consequences that may follow the dissemination of GMO crops and point to some of the unfortunate consequences of earlier applications of science to policy-making in the context of the Green Revolution of the 1970s–1980s (Dowie 1995). Their prescription is to make liberal use of the precautionary principle and to insist on high standards of proof of harmlessness in making decisions about products such as GMO crops.

In the eyes of some, risk analysis and risk management offer a way out of this impasse in the scientific world. The premise underlying this approach – widely used in areas such as public health – is that certainty is neither possible nor necessary as a basis for policy-making in situations of this kind. What is desirable is a careful assessment of possible consequences and the probability of their occurrence that makes it possible to arrive at objective judgements about the effects of specific decisions on social welfare. Whatever the merits of this approach (and there are those who question them) as applied to decisions about immunization programmes or the marketing of pharmaceuticals, there are good reasons to doubt whether procedures of this kind can help to resolve the differences between supporters of the WTO and proponents of the Biosafety Protocol. The problem is that neither the full range of possible consequences nor the relevant probabilities are sufficiently well known to provide a basis for carrying out risk assessments in this realm that all sides find persuasive, much less acceptable as a basis for policy-making. And, in the absence of credibility in scientific terms, risk analysis is more likely to exacerbate the politicization of arguments about GMO crops than to offer a viable method of reconciling the views of the two opposing camps in this debate.

Two other mechanisms are worthy of mention in this discussion of efforts to soften the confrontation between the divergent principles and norms underlying the WTO and the Protocol. One involves the creation of a Committee on Trade and Environment (CTE) as an element of the trade regime as expanded and modernized under the terms of the WTO adopted at the end of the Uruguay Round in 1994 (Shaffer 2002). Supporters of this addition to the trade regime point to the CTE as an effective response to the growing concerns about trade–environment tensions and cite evidence that this committee has played a constructive role in avoiding or alleviating frictions between the WTO and a number of important multilateral environmental agreements (e.g. the agreements relating to trade in endangered species, the depletion of stratospheric ozone, and transboundary movements of hazardous wastes). Whatever the merits of this claim, it is increasingly apparent that it will not satisfy the concerns of those who worry about dangers such as the long-term effects of GMO crops on human health and the resilience of ecosystems. In part, this is attributable to the fact that there is no getting around the status of the CTE as a component of the trade regime; it cannot be defended as a body that is equally receptive to the concerns of those on both sides of the divide between the WTO and the Protocol. But what makes the problem particularly severe in this case are the facts that there is no purely scientific way to resolve the core issues at stake and that the debate surrounding these issues has now become deeply politicized. In the final analysis, therefore, the CTE does not offer a mechanism that can alleviate problems arising from the interplay of the WTO and the Protocol, at least at this time.

In the absence of a substantive solution to the problems arising in this case, the drafters of the Protocol resorted to a number of well-known negotiating techniques intended to deflect attention from the divergences between the biosafety regime and the broader trade regime (Gupta 2000; Gupta, Chapter 2 in this volume). To begin with, they changed terminology in an attempt to soften the disparities between the WTO and the Protocol. Thus, the Protocol (Cartagena Protocol 2000) uses the term "living modified organism" (LMO) in place of "genetically modified organism" (GMO); it replaces the familiar idea of "prior informed consent" (PIC) with the new concept of "advance informed agreement" (AIA), and it refers to "intentional introduction" instead of "deliberate release" in describing the planting of GMO or LMO crops. Similarly, the preamble of the Cartagena Protocol states, at one and the same time, that the Protocol "shall not be interpreted as implying a change in the rights and obligations of a Party under any existing international agreements" and that "the above recital is not intended to subordinate this Protocol to other international agreements". What is more, the Protocol contains what is known as a "savings clause", which proclaims that the Protocol shall not affect the rights and obligations of parties under "any existing international agreement", except where this would cause serious damage or threat to biological diversity.

What should we make of these moves? It is well known that "constructive ambiguity" plays an important role in the realm of international governance and for that matter in governance more generally. It often makes sense to negotiate agreements containing ambiguities of this sort when the alternative is to admit defeat and to end up with no agreement at all. The hope, in such situations, is that differences can be resolved in practice through a process of mutual adjustment that allows two or more regimes to function effectively without interfering with one another at a deep level. In the case at hand, however, this justification seems unpersuasive. Although we cannot foresee clearly how relations between the WTO and the Protocol will play out in practice, the evidence available at this juncture suggests that these efforts to circumvent divergences have not gone far toward overcoming the differences in the principles, norms and values underlying the two regimes (Bernauer 2003).

Where does this leave us? If the gap between the underlying principles and norms of the WTO and the Biosafety Protocol were not large, coming to terms with this case of institutional interplay would not pose a major problem. In fact, it might be unnecessary even to have a separate protocol – much less a separate regime – to address concerns about biosafety. The introduction of suitable measures to allay concerns about human health and ecosystem resilience arising in conjunction with the development of biotechnology could be handled in a straightforward manner under the provisions of Article XX of the GATT and associated provisions in the SPS Agreement. But, clearly, such a move is not on the cards at this juncture. The gap between the trade regime and the biosafety regime is not narrowing. If anything, the forces that have created this gap are growing stronger as concerns about human domination of the Earth's ecosystems rise and the dominance of the post-war consensus regarding the value of a liberal international economic order becomes frayed at the edges (Bernauer 2003). Not only is the interplay between these regimes intentional; it is also deep and, for the most part, conflictual. It is accurate, in my judgement, to locate this case of interplay well into the lower right-hand quadrant of Figure 6.1, at least for now.

5. What is to be done?

Assuming that the argument of the preceding section is correct, how is the interplay between the WTO and the Cartagena Protocol likely to play out during the foreseeable future? Can we make any forecast concerning the evolution of this interaction over the next five to ten years? Can we recommend steps that could and should be taken to mitigate the consequences of this conflict, even if it is not possible to resolve it in some mutually acceptable fashion? In this section, I address these questions from two distinct perspectives. The first focuses on the politics of this case and asks whether an analysis of the interactions among the key players can shed light on the probable character of the next phase of the story of the WTO and the Protocol. The second perspective directs attention to issues of conflict management and resolution, asking whether it would be effective – or even desirable – to take steps now to defuse the conflict emerging from the interplay between the WTO and the Protocol.

5.1. The politics of the WTO and Cartagena Protocol case

Whether we like it or not, politics matters as a determinant of the outcomes of clashes between divergent principles, norms and values. Because power and influence are distributed in a highly uneven fashion in international society, a natural approach to answering the questions posed in the preceding paragraph is to examine the political fault lines associated with interplay between the WTO and the Protocol and to make projections about future developments based on an assessment of the distribution of bargaining strength in this setting (Odell 2000). Since the United States is the only remaining superpower in the world today and given the fact that the United States has expressed a clear preference for handling issues pertaining to trade in GMO/LMO crops within the framework of the WTO, a first-order prediction arising from such an analysis suggests that the prospects for successful implementation of the provisions of the Cartagena Protocol are poor.[4] The United States has refused to ratify the Convention on Biological Diversity, the larger arrangement in which the Protocol is embedded, much less to accept the terms of the Protocol itself as legally binding. How can the precautionary approach to trade in GMO/LMO products thrive in the face of this opposition on the part of the single most influential member of international society and the acknowledged world leader in the development of biotechnology (Bernauer 2003)?

What makes the politics of this case more complex – and more interesting – is the fact that concerns about the impacts of GMO/LMO products on human health and the resilience of ecosystems have given rise to major East–West differences as well as North–South differences (Andrée 2005). The interplay between the WTO and the Protocol involves three major blocs or negotiating groups:

- The Miami Group, an alliance of exporters of agricultural products led by the United States, advocates the adoption of rules relating to transboundary flows of all GMO/LMO products that will not impede the growth of trade in this area.
- The European Union, which has experienced growing public concern about matters such as food safety, has emerged as a supporter of a cautious approach to the acceptance of GMO/LMO products.
- The Like-Minded Group, which encompasses many but not all of the developing countries, includes players who are deeply concerned about the dangers associated with GMO/LMO products but who are, at the same time, sensitive to the potential benefits accruing from the use of these products when it comes to the issue of food security.

This broad characterization of the political landscape undoubtedly masks a good deal of complexity in the views of specific participants in the debates about biosafety (Gupta, Chapter 2 in this volume). Whereas food security is a critical issue for some participants, for instance, uncertainties about the long-term impacts of biotechnology on the preservation of biological diversity are a more pressing concern for others.

The differences between these blocs go to the heart of the clash between the principles, norms and values relating to international trade and biosafety. Not surprisingly, however, they also manifest themselves concretely in discussions of specific issues that have become vehicles for carrying on this debate. Prominent among these are differences relating to labelling, documentation and liability arrangements. And it is worth noting that the treatment of specific issues of this sort can affect the course of the larger debate in important ways.

Consider the issue of labelling in this light. Whereas the European Union has advocated strong rules and regulations governing the labelling of GMO/LMO products, the Miami Group under the leadership of the United States has resisted the adoption of such requirements. Whatever the ultimate merits of the arguments regarding this issue, the position of the Miami Group on the matter has not played well in the court of public opinion. For many, the value of labelling as a basis for truth in advertising and informed decision-making is so obvious that opposition to a strong position on the issue is suspect. On this score, the merits of the position of the Miami Group are hard to sell, whether or not they are justifiable in some technical sense.

As these observations suggest, moreover, the politics of this case is affected by broad trends in the evolution of public opinion. In this connection, there is little doubt that concerns about matters of public health and ecosystem resilience are growing in many quarters or that the traditional liberal support for free trade as an article of faith is showing signs of softness. Justifiably or not, both trust in the assurances of government agencies and belief in the objectivity of science are waning. Conversely, fear regarding threats to public health and suspicions of the motives of transnational corporations that are prominent advocates of biotechnology (e.g. Cargill, DuPont, Monsanto) are on the rise. It is hard to measure the effects of these larger and deeper trends in public sentiment, much less to compare the effects of this driving force with the impacts of more conventional sources of bargaining strength. Nonetheless, it would be unwise to disregard the influence of these broader forces working at the level of principles, norms and values, in contrast to mainstream perspectives on the dynamics of institutional bargaining.

Where are we headed with regard to institutional interplay between the WTO and the Protocol? The United States as the leader of the Miami Group has put pressure on the European Union with regard to this issue in the forum provided by the WTO.[5] It has also exerted steady pressure on the leaders of developing countries to accept foreign aid – including food aid – containing GMOs or LMOs. Moreover, there is a certain irreversibility when it comes to the spread of GMO/LMO products: once genetically modified organisms become part of an ecosystem, there is no guarantee that it will prove feasible to return to the status quo ante. On the other hand, opposition to the dissemination of GMO/LMO products is based on powerful feelings or sentiments that appear to be spreading through large parts of the world. Probably the safest prediction, in the circumstances, is that the clash associated with interplay between the WTO and the Protocol is destined to continue and very likely to become more intense during the next five to ten years. As a result, it seems pertinent at this stage to move on to a consideration of my second perspective and to ask about the relevance of ideas pertaining to conflict resolution and management to this case.

5.2. Resolving/managing the clash between the WTO and the Protocol

Perhaps the first thing to note in a discussion of this issue is that not all conflict is bad. It is true, of course, that conflict – like bargaining or negotiation – can be costly. It is perfectly reasonable to treat the time, energy and material resources that participants invest in the prosecution of conflicts as transaction costs or as opportunity costs in the sense that resources consumed in conflicts are not available for other purposes. Nonetheless, as sociologists of conflict from Simmel to Coser have pointed out, conflict can produce benefits as well (Coser 1956). Conflict can sharpen and deepen our understanding of complex issues; it can be an energizing force; and, handled properly, it can contribute to the emergence of outcomes that are fair or just. The philosophy underlying the advocacy process as a means of resolving legal disputes, for example, emphasizes the social benefits of conflict, so long as it takes place in a controlled setting subject to the rule of law (Fuller 1981; Auerbach 1983).

Coupled with the argument of section 5.1 about the politics of the interplay between the WTO and the Protocol, these observations suggest the importance of examining options for managing rather than resolving the clash between the WTO and the Protocol, at least during the foreseeable future. We can say at once that technical solutions (e.g. giving priority to agreements that are more recent in time or that contain more precise provisions) will not suffice to keep this case of institutional interplay under control. The problem arises from a clash of principles, norms and values, coupled with a high degree of uncertainty about the longer-term impacts of various forms of biotechnology. Technical advances may prove helpful, even in such situations. Developing new and better methods for tracking GMO/LMO products through space and time, for instance, would certainly make it easier to implement regulations relating to the documentation of trade in such products and to monitoring the impacts of such products. But technical advances of this sort cannot alleviate more deeply rooted concerns about the longer-term consequences of GMO/LMO products for human health and the resilience of ecosystems.

Any effort to manage this clash must take into account several factors that arise from the nature of this particular case as well as from the character of international society more broadly. We can say with high certainty that mechanisms attached to or embedded in one regime cannot succeed in managing – much less resolving – conflicts arising from interplay between regimes that is intended, deep and, for the most part, conflictual. The obvious case in point is the Committee on Trade and Environment (CTE) established as a component of the WTO in the package of agreements adopted at the end of the Uruguay Round in 1994. Whatever claims are made on behalf of the CTE and however well it performs in specific situations, there is no way to allay the fears of those worried about matters of biosafety regarding the institutional biases of the CTE. What is needed in this connection is an unbiased mechanism with both the authority and the legitimacy to render judgments in cases involving tensions between the two regimes. In well-established democracies, the courts normally play an important role in meeting this functional need; that is why the appointment of judges to higher-level courts is such a sensitive matter. But this is exactly what international society as a whole lacks. Where technical solutions will suffice, it is possible to rely on a number of mechanisms – including the United Nations and the various specialized agencies – to play this role. However, this procedure will not work as a method of dealing with frictions arising from interplay that is intended and deep.

We do know that meaningful stakeholder participation is beneficial to efforts to resolve or manage conflicts in a variety of circumstances. Parties are willing to accept a remarkably wide range of outcomes if they are convinced that the process of arriving at them is fair and that they were accorded a proper role in arriving at specific conclusions. But how can this work in this particular case, where many regard the trade side as relatively closed, élitist and self-centred and the environmental side as militant and doctrinaire? Whatever the role of stakeholders in the politics of this case, more is needed than public confrontations of the sort that occurred in Seattle in 1999 and the subsequent efforts of the parties to the WTO to avoid a repeat of Seattle by meeting in Doha. It is certainly possible to imagine leaders of the two movements – the international trade movement and the environmental movement – meeting to discuss their differences in a neutral setting. In practice, however, this is easier said than done. In most member states, there is little interaction at the domestic level regarding tensions of the type exemplified by this case. Trade officials and their environmental counterparts work in their own units, develop positions on issues without extensive interactions with their counterparts in other units, and field delegations at international meetings that are rather narrow. The result is a bifurcated process in which member states interact with one another in a functionally segmented manner. A first step toward managing interplay of the sort exemplified by the WTO and the Protocol case, then, would be for the governments of key states to initiate a dialogue among those located in different agencies at the domestic level on the critical issues. This would not guarantee success at the international level, but it would certainly help to move us toward a situation in which the key stakeholders are able to engage in a common discourse rather than triggering open confrontations of the sort that occurred in Seattle in 1999.

Beyond this, any successful effort to manage tensions arising out of interplay between the WTO and the Protocol must address the need to find an appropriate mode of participation for non-state actors, including civil society organizations as well as transnational corporations (TNCs). There is no implication here that the role of states is waning in this realm, or for that matter more generally. But TNCs are critical players in the world of trade, and environmental organizations have assumed a leading role in developing and articulating the case for proceeding with caution when it comes to trade in GMO/LMO products (Smith et al. 1997; Khagram et al. 2002). Just as there is a problem when functionally specific agencies within governments do not engage in a constructive dialogue on matters of this sort, managing the tensions arising from the interactions between the WTO and the Protocol will require some sort of engagement between the TNCs and major environmental organizations.

Unfortunately, it is not likely that the scientific community can offer much help in this case. Scientists themselves are divided between a mainstream group associated with organizations such as the Food and Agriculture Organization of the United Nations, the Codex Alimentarius Commission and the Consultative Group on International Agricultural Research, whose members tend to assume that it is perfectly possible for scientists to make sound judgements about matters such as biosafety, and a more critical group associated with programmes such as the Millennium Ecosystems Assessment, whose members are more sensitive to the dangers arising from the loss of biological diversity and the uncertainties regarding the longer-term impacts of biotechnology on the resilience of ecosystems. The resultant picture is complex; it is far from clear how to provide an appropriate voice for non-state actors in dealing with the interplay between the WTO and the Protocol. Perhaps there is a role here for a high-level but autonomous body – analogous to the World Commission on Dams (World Commission on Dams 2000) – that can sort through the principal issues surrounding this case and articulate a framework and a set of criteria for managing the tensions embedded in the interactions between the WTO and the Protocol.

To put all this in perspective, it is important to remember that the tensions under consideration here are not first-order issues at the level of international society. Thus, these tensions are not likely to precipitate violent conflicts; they do not pose – in the language of the UN Charter – "threats to the peace, breaches of the peace, [or] acts of aggression" (Schlesinger 2003).[6] That is the good news. Nevertheless, the tensions in question are deeply rooted in divergent principles, norms and values; they are not going to go away during the foreseeable future. If anything, they will become more severe as concerns about the impact of anthropogenic drivers on planetary support systems rise and pressures to avoid dangerous interventions in these systems become more sharply focused. It will not do, therefore, simply to let the tensions between the WTO and the Protocol play themselves out on their own and hope for the best. There is a need – as a matter of priority – to launch a concerted effort to manage the tensions associated with this case of institutional interplay.

Those desiring to downplay the significance of this conflict or to put off making a concerted effort to come to grips with the underlying issues can take comfort in a number of compromises agreed to by those who negotiated the text of the Cartagena Protocol (Gupta, Chapter 2 in this volume; Oberthür and Gehring, Chapter 5 in this volume). Both the preamble of the Protocol and several of its substantive provisions were drafted with care to create at least an illusion of compatibility with the provisions of the WTO. The Protocol subjects GMOs/LMOs intended for food, feed and processing to different rules from those for GMO/LMO crops. As Oberthür and Gehring suggest in this volume, these provisions could strengthen initiatives within the WTO aiming at a "peaceful coexistence" of both regimes because of the interest of parties to the Cartagena Protocol in avoiding incompatible commitments. One interesting idea along these lines would be to give those associated with the Protocol some acknowledged role in standard-setting under the terms of the SPS Agreement.

Some participants may have distinct incentives to follow such a strategy, adopting a shallow perspective on the interplay between these regimes and keeping the deeper issues off the table to the extent possible. This strategy might work, at least in the short run. It seems unlikely, however, to yield a permanent solution to the tensions embedded in the interaction between the WTO and the Cartagena Protocol. If, as I believe is likely, the forces questioning the neo-liberal orthodoxy underlying the WTO and turning to issues of food security as one basis for raising doubts about the consequences of the established world trading system gain momentum over the next five to ten years, the deeper issues embedded in this interplay will inevitably come to the fore. There is no reason to expect that they will become easier to deal with – much less go away – as a result of being glossed over in the short run.

6. A note on methodology

The analysis presented in this chapter has an important methodological implication that is worth pausing to comment on at this juncture. Prior work on institutional interplay has fluctuated between the poles of what I would call reductionism and integrationism. The reductionist approach, exemplified in the work of Oberthür and Gehring (2006), takes the view that we need to dissect interplay, focusing first on the simplest possible case, in which there are two distinct regimes – a source and a target – and a single, unidirectional flow of influence going from one to the other. A prominent example involves the influence that the Montreal Protocol on ozone depletion has had on the climate regime by curtailing the production and consumption of ozone-depleting chemicals that also turn out to be greenhouse gases. Of course, it is possible to build up a more complex assessment of interplay both by considering additional flows of influence and by widening the coverage of the analysis in institutional terms. Even so, the method is essentially additive; it proceeds to develop a fuller picture by focusing on the discrete unidirectional flow of influence as the unit of analysis and adding more of these units to the analysis until the picture is complete.

The integrationist approach, exemplified by the recent work of Raustiala and Victor on plant genetic resources, introduces the idea of regime complexes treated as interlocking sets of institutional arrangements and takes the view that interplay manifests itself in the form of emergent properties of these complexes or institutional systems (Raustiala and Victor 2004). The case they use to explore this idea encompasses the International Convention for the Protection of New Varieties of Plants, the International Treaty on Plant Genetic Resources, the Consultative Group on International Agricultural Research, the WTO's TRIPs Agreement and the Convention on Biological Diversity. In analysing this case, Raustiala and Victor develop a number of specific conjectures about the dynamics of regime complexes. But their central claim is that "regime complexes evolve in ways that are distinct from decomposable single regimes" (Raustiala and Victor 2004: 279).

The case of the WTO and the Cartagena Protocol, by contrast, suggests a somewhat different perspective on the nature of institutional interplay as well as the appropriate methods for examining it. In this case, interplay is essentially relational. It is not a matter of unidirectional influence flowing from the WTO to the Protocol, or vice versa. Nor is it a matter of unplanned interactions among a number of institutional arrangements dealing with the same general issue area. Rather, it arises from the fact that the two regimes reflect different approaches to an activity or problem that can be addressed in a number of ways. From one perspective, trade in GMO/LMO products is simply a particular form of international trade that could be handled under the general rules of the WTO, including the SPS Agreement. From another, it is a distinct activity that raises unique issues that can and should be subjected to a set of rules that are *sui generis* and that are designed to meet the special problems of this type of trade. However we choose to look at it, the problem is not a matter of unidirectional influence or of the emergent properties of a regime complex; it involves a divergence of views regarding the proper approach to governing a particular issue area.

What is the significance of these methodological concerns, and how should we respond to them? In my judgement, the crux of this issue lies in the distinction between unintended and intended interplay. Where interplay is a matter of unintended side effects or externalities, it makes sense to begin with the unidirectional flow as the basic unit of analysis. There may well be cases in which flows of this kind go in both directions. A regime that protects tropical forests in the interests of preserving biological diversity, for example, will have the effect of limiting emissions of greenhouse gases into the Earth's atmosphere. For its part, a regime that protects the Earth's climate system from dramatic, anthropogenic shifts can be expected to prove beneficial for the health of ecosystems and, therefore, for the preservation of biological diversity. But this does not alter the argument for adopting the unidirectional flow of influence as the unit of analysis. Similar remarks are in order regarding interactions among the components of regime complexes. So long as the interactions in question are unintended side effects of actions taken to fulfil the objectives of individual components, the relational phenomenon described in the preceding paragraph does not arise.

Once we move into the realm of intended interactions, on the other hand, interplay becomes relational. The creators of a new regime (e.g. the Cartagena Protocol) are motivated, at least in part, by the fact that they do not want to see an existing regime extended to cover new concerns. On the contrary, they want to undermine the influence of an existing regime with regard to some particular issue(s). Those responsible for operating the existing regime, by contrast, will seek to extend it to the new activity (e.g. trade in GMO/LMO products) or to defend the regime from inroads arising from the operation of the new regime. Of course, the depth of the relevant interplay will become important as well in such matters. In brief, the deeper the interplay, the more significant the relational character of the resultant interactions becomes. I conclude from these observations that there is room for examining institutional interplay in reductionist, integrative and relational terms. The trick is to make clear and informed choices about the proper method to be used in individual cases.

7. Concluding remarks

I have argued that, although institutional interplay is an important topic and a fruitful area for research on international governance, our initial forays into this field of study leave a good deal to be desired. To put it bluntly, we have shown a propensity to think in taxonomic terms without paying adequate attention to the theoretical significance of the distinctions we introduce. The remedy I propose is to cut back drastically on the development of new concepts relating to interplay and to spell out and examine critically the theoretical significance of the distinctions we do emphasize. I use the interplay between the WTO and the Protocol regarding transboundary flows of GMO/LMO products as an empirical test bed for exploring these claims.

Specifically, I have argued that this case features interplay that is intended, deep and conflictual in nature. We should expect that interplay of this sort will be the hardest to come to terms with effectively. Intended interplay that is synergistic in character, by contrast, should be comparatively easy to address. It can be expected to produce opportunities for mutual gains and outcomes that are stable in the sense that none of the participants has any incentive to cheat once an agreement on how to proceed is reached. Unintended interplay should occupy an intermediate position on this scale. Those responsible for negative externalities in the realm of institutional interplay will have no natural incentive to alter their practices; beneficiaries of positive institutional externalities will see little or no reason to pay for them or to invest resources in enhancing their constructive consequences. In general, the evidence from this case is compatible with these expectations. But, of course, it is essential to avoid drawing unwarranted conclusions from a single case. What is needed now are systematic efforts to refine expectations derived by theory of the sort discussed in this chapter, through systematic analyses of additional cases rather than additional efforts to construct conceptual maps of the various dimensions of interplay. The study of institutional interplay is both scientifically challenging and relevant to policy-making relating to many issues. We have gained some important insights into this phenomenon already, but the field is wide open for cutting-edge research on the part of a new generation of analysts.

Notes

1. Although it is not formally part of the WTO, the Codex Alimentarius is closely integrated into the trade regime in functional terms.
2. In addition to the approaches represented by the chapters of this volume, see the work of Andersen (2002), Andrée (2005), Bail et al. (2002), Falkner (2002), Gupta (2000), Oberthür (2001), Oberthür and Gehring (2006), Rosendal (2001), Safrin (2002), Selin and VanDeveer (2003), Steinberg (2002), Stokke (2001) and Underdal and Young (2004).
3. For clear examples relating to whales and whaling, see Friedheim (2001).
4. In what is clearly a bow to political realities, the Protocol does not cover pharmaceuticals and places fewer restrictions on trade in LMOs intended for food, feed and processing than on trade in LMOs intended for use as crops (Gupta, Chapter 2 in this volume).
5. For instance, the United States has a case pending in the WTO against EU regulations relating to transgenic crops (Gupta, Chapter 2 in this volume).
6. The phrase in quotes is from the title of Chapter VII of the UN Charter.

REFERENCES

Andersen, Regine (2002), "The Time Dimension in International Regime Interplay", *Global Environmental Politics* 2(3): 98–117.
Andrée, Peter (2005), "The Genetic Engineering Revolution in Agriculture and Food: Strategies of the 'Biotech Bloc'", in David L. Levy and Peter J. Newell

(eds), *The Business of Global Environmental Governance*. Cambridge, MA: MIT Press, pp. 135–166.

Andresen, Steinar et al. (2000), *Science and Politics in International Environmental Regimes: Between Integrity and Involvement*. Manchester: Manchester University Press.

Auerbach, Jerold S. (1983), *Justice without Law: Resolving Disputes without Lawyers*. New York: Oxford University Press.

Bail, Christoph, Robert Falkner and Helen Marquard, eds (2002), *The Cartagena Protocol on Biosafety: Reconciling Trade in Biotechnology with Environment and Development?* London: Earthscan.

Bernauer, Thomas (2003), *Genes, Trade, and Regulation: The Seeds of Conflict in Food Biotechnology*. Princeton, NJ: Princeton University Press.

Cartagena Protocol (2000), *Cartagena Protocol on Biosafety to the Convention on Biological Diversity: Text and Annexes*. Montreal: Secretariat of the Convention on Biological Diversity; available at ⟨http://www.cbd.int/doc/legal/cartagena-protocol-en.pdf⟩ (accessed 5 July 2007).

Coser, Lewis A. (1956), *The Functions of Social Conflict*. Glencoe, IL: Free Press.

Dowie, Mark (1995), *Losing Ground: American Environmentalism at the Close of the Twentieth Century*. Cambridge, MA: MIT Press.

Falkner, Robert (2002), "Negotiating the Biosafety Protocol: The International Process", in Christoph Bail, Robert Falkner and Helen Marquard (eds), *The Cartagena Protocol on Biosafety: Reconciling Trade in Biotechnology with Environment and Development?* London: Earthscan, pp. 3–22.

Friedheim, Robert L., ed. (2001), *Toward a Sustainable Whaling Regime*. Seattle: University of Washington Press.

Fuller, Lon L. (1981), *The Principles of Social Order*. Durham, NC: Duke University Press.

Gilpin, Robert (2001), *Global Political Economy: Understanding the International Economic Order*. Princeton, NJ: Princeton University Press.

Gupta, Aarti (2000), "Governing Trade in Genetically Modified Organisms: The Cartagena Protocol on Biosafety", *Environment* 42(4): 22–23.

Irwin, Douglas A. (2002), *Free Trade under Fire*. Princeton, NJ: Princeton University Press.

Khagram, Sanjeev, James V. Riker and Kathryn Sikkink, eds (2002), *Restructuring World Politics: Transnational Social Movements, Networks, and Norms*. Minneapolis: University of Minnesota Press.

Lang, Tim and Colin Hines (1993), *The New Protectionism: Protecting the Future against Free Trade*. New York: New Press.

Mishan, E. J. (1982), *What Political Economy Is All About: An Exposition and Critique*. Cambridge: Cambridge University Press.

Oberthür, Sebastian (2001), "Linkages between the Montreal and Kyoto Protocols: Enhancing Synergies between Protecting the Ozone Layer and the Global Climate", *International Environmental Agreements: Politics, Law and Economics* 1(3): 357–377.

Oberthür, Sebastian and Thomas Gehring, eds (2006), *Institutional Interaction: Synergy and Conflict between International and EU Environmental Regulations*. Cambridge, MA: MIT Press.

Odell, John S. (2000), *Negotiating the World Economy*. Ithaca, NY: Cornell University Press.
Ostrom, Elinor et al., eds (2002), *The Drama of the Commons*. Washington, DC: National Academy Press.
Pollan, Michael (2001), *The Botany of Desire: A Plant's Eye View of the World*. New York: Random House.
Putnam, Robert D. (1988), "Diplomacy and Domestic Politics: The Logic of Two-Level Games", *International Organization* 42: 427–460.
Raustiala, Kal and David G. Victor (2004), "The Regime Complex for Plant Genetic Resources", *International Organization* 58(2): 277–309.
Rosendal, Kristin (2001), "Impacts of Overlapping International Regimes: The Case of Biodiversity", *Global Governance* 7(1): 95–117.
Ruggie, John Gerard (1983), "International Regimes, Transactions, and Change: Embedded Liberalism in the Postwar Economic Order", in Stephen D. Krasner (ed.), *International Regimes*. Ithaca, NY: Cornell University Press, pp. 195–231.
Ruttan, Vernon W., ed. (1994), *Agriculture, Environment, and Health: Sustainable Development in the 21st Century*. Minneapolis: University of Minnesota Press.
Safrin, Sabrina (2002), "Treaties in Collision? The Biosafety Protocol and the World Trade Organization Agreements", *American Journal of International Law* 96(3): 606–627.
Schlesinger, Stephen C. (2003), *Act of Creation: The Founding of the United Nations*. Boulder, CO: Westview Press.
Selin, Henrik and Stacy D. VanDeveer (2003), "Mapping Institutional Linkages in European Air Pollution Politics", *Global Environmental Politics* 3: 14–46.
Shaffer, Gregory C. (2002), "The Nexus of Law and Politics: The WTO's Committee on Trade and Environment", in Richard H. Steinberg (ed.), *The Greening of Trade Law: International Trade Organizations and Environmental Issues*. London: Rowman & Littlefield, pp. 81–114.
Smith, Jackie, Charles Chatfield and Ron Pagnucco, eds (1997), *Transnational Social Movements and Global Politics: Solidarity Beyond the State*. Syracuse: Syracuse University Press.
Stein, Arthur A. (1983), "Coordination and Collaboration Regimes in an Anarchic World", in Stephen D. Krasner (ed.), *International Regimes*. Ithaca, NY: Cornell University Press, pp. 115–140.
Steinberg, Richard H., ed. (2002), *The Greening of Trade Law: International Trade Organizations and Environmental Issues*. London: Rowman & Littlefield.
Stokke, Olav S. (2001), "The Interplay of International Regimes: Putting Effectiveness Theory to Work", FNI Report 14, Fridtjof Nansen Institute, Oslo.
Tietenberg, Thomas (2002), "The Tradable Permits Approach to Protecting the Commons: What Have We Learned?", in Elinor Ostrom et al. (eds), *The Drama of the Commons*. Washington, DC: National Academy Press.
Underdal, Arild and Oran R. Young, eds (2004), *Regime Consequences: Methodological Challenges and Research Strategies*. Dordrecht: Kluwer Academic Publishers.
Vitousek, Peter, Harold Mooney, Jane Lubchenko and Jerry Melillo (1997), "Human Domination of the Earth's Ecosystems", *Science* 277: 494–499.

World Commission on Dams (2000), *Dams and Development: The Report of the World Commission on Dams*. London: Earthscan.

Young, Oran R. (2006), "Vertical Interplay among Scale-Dependent Environmental and Resource Regimes", *Ecology and Society* 11(1): 27 [online], www.ecologyandsociety.org/vol11/iss1/art27/.

Young, Oran R., with contributions from Arun Agrawal, Leslie A. King, Peter H. Sand, Arild Underdal and Merrilyn Wasson (1999), *Science Plan: Institutional Dimensions of Global Environmental Change*, IHDP Report No. 9. Bonn: International Human Dimensions Programme on Global Environmental Change.

Part IV
Remembering Konrad von Moltke

Teil IX

Helmuth Konrad von Moltke

7

The WTO as an environmental agency

Steve Charnovitz

1. Introduction

This chapter provides an overview of the "trade and environment" issue in the World Trade Organization (WTO) and recent developments in related WTO jurisprudence. My study was prepared as part of a research project organized by United Nations University to examine the interplay of international trade and biosafety with special reference to the new Cartagena Protocol on Biosafety. Recently, an article in the *Journal of World Trade* criticized the Protocol as "a club for agricultural protectionists" (Hobbs et al. 2005: 297), and it remains to be seen how a governmental decision taken pursuant to the Protocol would fare if challenged in WTO dispute settlement.

The editors of this volume made a wise choice in commissioning this chapter from Konrad von Moltke, one of the most respected and popular analysts of international environmental policy during the past quarter-century. Back in 1990, when I first began writing about the intersection of trade and environment, Professor von Moltke was one of the few scholars in the world to whom one could turn for guidance, because he had already given considerable thought to the looming clash. He was happy to tutor me, and soon became a good friend. Coming as I did from the trade side of the debate, Konrad seemed to relish the opportunity to explain to me how to integrate environmental analysis into a trade perspective. After he tragically passed away in May 2005, I joined others in a global email conversation to lament this loss.

Institutional interplay: Biosafety and trade, Young, Chambers, Kim and ten Have (eds), United Nations University Press, 2008, ISBN 978-92-808-1148-3

When the editors of this volume asked me to substitute for von Moltke in writing this chapter, I recalled his inimitable style and his important papers on trade and environment (e.g. von Moltke 1993, 1996). Readers of von Moltke were always treated to a fresh and integrative approach to any new issue he tackled. We also gained from his ability to think out of the box and put forward a provocative thesis that would cause readers to rethink their assumptions. Inspired by Konrad's example, I offer a daring thesis here – that we should visualize the WTO as an environmental agency.

The chapter is structured as follows. Section 2 provides a brief review of the history of the environment linkage in trade policy, beginning in 1923. Section 3 presents my thesis that the WTO should have a positive environmental role. Section 4 looks at the many ways that the environment already features in WTO rules and case law. It also provides an overview of how trade rules may hinder environmental policy. Section 5 looks at the environmental components of the WTO's Doha Round negotiations. Section 6 presents the concept of the multifunctional international organization and explains why the traditional paradigm of the WTO as a trade-only agency needs to be replaced by a new paradigm. Section 7 concludes.

2. Background on the trade–environment linkage

At its origins in the 1920s, the trading system sought to avoid interfering with national health and environmental policy measures. The first multilateral treaty on trade, the Convention Relating to the Simplification of Custom Formalities of 1923, contained a provision stating that the disciplines of the treaty did not "prejudice the measures which contracting parties may take to ensure the health of human beings, animals or plants" (Customs Simplification Convention 1923: Article 17). The next major treaty was the Convention for the Abolition of Import and Export Prohibitions and Restrictions of 1927. The drafter of the Convention wrote in an exception for "prohibitions or restrictions imposed for the protection of human health and for the protection of animals and plants against disease, insects and harmful parasites" (Trade Prohibitions Convention 1927: Article 4.4). After the treaty was negotiated, there was some concern about whether this exception was sufficiently capacious. Therefore, a Protocol was added to clarify that this exception "also refers to measures taken to preserve them [animals and plants] from degeneration or extinction and to measures taken against harmful seeds, plants, parasites and animals" (Trade Prohibitions Convention 1927: Protocol, ad Article 4(a)). The Protocol makes clear that, even by 1927, governments were thinking about the repercussions of international trade rules on biodiversity and biosafety.

When governments negotiated the General Agreement on Tariffs and Trade (GATT) and the Charter of the International Trade Organization (the Havana Charter), 20 years later, there were a sufficient number of multilateral environmental agreements in place with specific trade objectives that the treaty drafters took care to add a general exception for measures "taken in pursuance of any inter-governmental agreement which relates solely to the conservation of fishery resources, migratory birds or wild animals" (Havana Charter 1948: Article 45(1)(a)(x)). The immediate post–World War II period had been an active time for international environmental policy-making, with the negotiation of the Whaling Convention of 1946, the Fishing Nets Convention of 1946, the Pan American Nature Protection Convention of 1948, and the constitutive act of the International Union for the Conservation of Nature and Natural Resources of 1948.

Unfortunately, the Havana Charter failed to come into force. In its place, the GATT remained the fundamental law of the trading system until the WTO came into being in 1995.

The GATT had little involvement with environmental issues until the early 1970s. In 1971, the GATT Secretariat prepared a report on "Industrial Pollution Control and International Trade" as an intellectual contribution to the forthcoming United Nations Conference on the Human Environment. Also that year, the GATT established a standby Group on Environmental Measures and International Trade. In addition, the GATT Secretariat gave technical advice to the drafters of the Convention on International Trade in Endangered Species of Wild Fauna and Flora (CITES) on how to make its trade obligations GATT consistent (Boardman 1981: 89–92).

The GATT Group took 20 years to hold its first meeting, and that occurred following a growing chorus of public concern that the GATT might be acting in an environmentally blind way. The Group met intermittently over the next couple of years until it was replaced in 1995 by the WTO Committee on Trade and Environment (CTE). In the years since the issue of the environment returned to the GATT in 1990, one can see that the efforts in the GATT and the WTO to consider environmental linkages have contributed to a better understanding of those challenges and to better coordination of policy-making at the national level (Shaffer 2001).

The scholarly output on "trade and the environment" is extensive and includes contributions from lawyers, economists, international relations specialists and scientists.[1] In this short chapter, I will not try to summarize that literature or to detail the many ways in which trade flows affect the environment[2] or in which environmental measures may restrict trade. Instead, I move directly to a new thesis about the WTO's role.

3. The case for a WTO environmental role

In its 2004 pamphlet entitled *Trade and Environment at the WTO*, the WTO Secretariat declares that one of the "parameters" for WTO discussion of trade and environment is that the "WTO is not an Environmental Protection Agency" (WTO 2004: 6). The Secretariat may be right that such a proposition underlies current thinking inside the WTO. Nevertheless, I doubt the accuracy of the proposition itself. In some ways, today's WTO is already an environmental agency and is becoming more of one.

My new thesis cuts against the grain. The traditional thinking is that the WTO is a trade liberalization agency and its success in performing that mission depends on maintaining its distinctive function. Many officials at the GATT and the WTO have sought to reassure environmentalists worried about the expanding reach of the trading system that the WTO has no interest in setting environmental rules or in the competence to do so. Along those lines, the Uruguay Round "Decision on Trade and Environment" asserts that the "competence of the multilateral trading system" is "limited to trade policies and those trade-related aspects of environmental policies which may result in significant trade effects for its members" (WTO 1994a).

The WTO has been colourfully described over the past decade, but I do not recall anyone giving it the appellation of "environmental agency". An excellent volume on the WTO published in 1998 was titled *The WTO as an International Organization* (Krueger 1998). Yet even that volume, edited by the eminent free-trader Anne Krueger, contained chapters on non-traditional topics such as "domestic political objectives" and "environmental and labour standards".

In considering whether the WTO is or is not an environmental agency, one should reflect on what it means to be an environmental agency (or an environmental protection agency). In my view, an environment agency is an agency that (1) makes assessments of environmental needs; (2) decides the level of environmental protection to be sought; or (3) selects the appropriate measures for achieving it.

By that definition, the WTO is an environmental agency.[3] Its scope of oversight potentially includes any governmental environmental measure (of a WTO member country) that affects trade. Under current rules, the WTO is certainly engaged in the third task and can perform the second to the extent that it requires countries to use international standards. The

WTO is not currently making assessments of environmental needs, but this could arise in the Doha Round negotiations.

In calling the WTO an environmental agency, I am not suggesting that such a descriptor is the best one for the WTO. The beginning of wisdom is to recognize that the WTO is multifunctional. It is primarily a trade liberalization agency, but it also plays an overlapping role in many regimes. As noted in the 2003 "Final Declaration" of the Parliamentary Conference on the WTO, organized jointly by the Inter-Parliamentary Union and the European Parliament, the "WTO is rapidly becoming more than a mere trade organisation" (Parliamentary Conference on the WTO 2003: para. 8).

Besides being a trade liberalization agency, the WTO has taken on additional identities. The WTO is an agriculture agency that addresses food aid (Zhang 2004). The WTO is an intellectual property agency.[4] Since the Doha Ministerial Conference of 2001, the WTO has become a development agency too.[5]

In calling the WTO an environmental agency, I am placing the WTO within the rather large population of international environmental agencies. Indeed, the fragmented nature of world environmental governance has become a serious problem and one in need of organizational reform (von Moltke 2005). Besides the WTO, the World Bank is another major multifunctional agency with an environmental mission.

How essential is my thesis to this study? For much of the descriptive and analytical material to be presented below, my thesis is not critical. The WTO will be a conditioning factor in biosafety policy whether or not one views the WTO as an environmental agency. Where my thesis is critical is in the discussion of how better to integrate trade and environmental law and how to transform the WTO into a pro-environment agency.

My thesis would be objected to by many. Some analysts take the view that the WTO should be only a market access agency. The economist Robert Staiger has taken that position in his thoughtful scholarship on the WTO. Staiger would be the first to acknowledge that the WTO of today has strayed from that singular mission, and he recommends disentangling trade from other issues and refocusing it on "securing market access property rights" (Staiger 2004: 13).

Yet if the WTO is exclusively a market access property rights agency, aloof from the environment regime, then that separate positioning facilitates the erroneous view that trade law is superior to environmental law.[6] The danger in allowing the WTO to view itself as outside the environment regime is that the WTO can just say "no" to a national environment or public health measure without taking any responsibility for the repercussions of its decision and, when warranted, getting the parties to a "yes". International governance can be dysfunctional when negative decisions may be taken in one international organization without any connection to whether positive decisions are taken in a parallel organization.

For 20 years or so, the paradigm for how the trading system interacts with environmental (and other "non-trade") issues has been "linkage" (see Alvarez 2002). Analysts have focused on the policy tensions that develop when the trade regime pursuing its own objectives crosses paths with the environment regime pursuing its own objectives.[7] The underlying assumption in the linkage paradigm is that the trading system is about trade, not about environment, and so environmental claims can enter only via linkage. Yet for many governments and stakeholders in the trade community, linkage is a dirty word and not one that is gaining in popularity.

The time has come to escape from the mental imprisonment of linkage. Back in 1992, the governments drafting *Agenda 21* for the United Nations Conference on Environment and Development (UNCED) stated that the "international community should: ... Ensure that environment and trade policies are mutually supportive, with a view to achieving sustainable development" (UNCED 1992: para. 2.10(d)). This notion of mutual supportiveness has been repeated in other intergovernmental statements and yet, even some 15 years later, governments have not made much progress in thinking through what it means for trade policy to be mutually supportive with environmental policy (and vice versa). Over the years, excellent books have been written about "The Greening of World Trade Issues", "Greening the GATT" and "The Greening of World Trade Law". In section 4 I consider how much the trading system has been greened.

4. The environment in WTO law

The WTO's attention to the environment starts at the beginning of the WTO treaty. In the Preamble to the Agreement Establishing the WTO, the parties act to establish the WTO,

> Recognizing that their relations in the field of trade and economic endeavour should be conducted with a view to raising standards of living,... while allowing for the optimal use of the world's resources in accordance with the objective of sustainable development, seeking both to protect and preserve the environment and to enhance the means for doing so in a manner consistent

with their respective needs and concerns at different levels of economic development. (WTO 1994b: Preamble)

In the *Shrimp* case, in 1998, the Appellate Body drew attention to this provision and used it to help interpret the general exceptions in GATT Article XX. The appellators famously stated that the Preamble "informs" all of the WTO trade agreements and "explicitly acknowledges 'the objective of *sustainable development*'".[8] In reference to this and other language in *Shrimp*, Professor John Jackson calls that decision "a constitutional door opener for approaches that require a broader perspective than just the four corners of the very extensive GATT/WTO treaty language" (Jackson 2005: 40).

Because of the controversy surrounding the *Shrimp* case and the fact that the jurists ruled against the US conservation measure being challenged, the Appellate Body included a coda at the end of its holdings to underscore what it had not decided. According to the Appellate Body,

> We have *not* decided that the protection and preservation of the environment is of no significance to the Members of the WTO. Clearly, it is. We have *not* decided that the sovereign nations that are Members of the WTO cannot adopt effective measures to protect endangered species, such as sea turtles. Clearly, they can and should. And we have *not* decided that sovereign states should not act together bilaterally, plurilaterally or multilaterally, either within the WTO or in other international fora, to protect endangered species or to otherwise protect the environment. Clearly, they should and do.[9]

Two features of this holding should be noted. First, the Appellate Body declares that states "should" adopt effective measures to protect endangered species. Perhaps that statement can be written off as a rhetorical flourish. Second, the Appellate Body seems to be suggesting that states can and perhaps should act together plurilaterally or multilaterally *within the WTO* to protect endangered species or otherwise to protect the environment. That statement is harder to overlook. It has to reflect an assumption by the Appellate Body that such collective action within the WTO would be consistent with the WTO's competence.

In the follow-up compliance dispute in *Shrimp*, the Panel held that "sustainable development is one of the objectives of the WTO Agreement".[10] This remarkable statement drew no criticism when the WTO Dispute Settlement Body adopted the Panel decision. To be sure, there is a difference between a holding that "sustainable development" is a WTO objective and a holding that environmental protection is a WTO objective. Yet, had I limited my thesis in this chapter to a proposition that the WTO is a sustainable development agency, that too would have been a major departure from the conventional view that the WTO is merely a trade agency.[11]

In its 1998 *Shrimp* ruling, the Appellate Body took note of the Uruguay Round "Decision on Trade and Environment", and held that this Decision can "help to elucidate the objectives of WTO Members with respect to the relationship between trade and the environment".[12] In particular, the Appellate Body quoted from the terms of reference for the Committee on Trade and Environment, which include whether to make recommendations for modifications of WTO provisions as regards, in particular,

- the need for rules to enhance positive interaction between trade and environmental measures, for the promotion of sustainable development, with special consideration to the needs of developing countries, in particular those of the least developed among them; and
- the avoidance of protectionist trade measures, and the adherence to effective multilateral disciplines to ensure responsiveness of the multilateral trading system to environmental objectives set forth in *Agenda 21* and the Rio Declaration, in particular Principle 12. (WTO 1994a)

This mandate admits of more than one interpretation. At the very least, it shows that governments agreed to assess whether the WTO should have provisions to achieve *positive* interaction between trade and the environment, to promote sustainable development and to ensure WTO responsiveness to international environmental objectives. A more expansive view is that WTO rules already promote those goals and the issue to be decided is whether those rules should be enhanced. So far, the Committee has not made decisions either way.

The WTO comprises 24 covered agreements and other understandings that are part of a single undertaking. Many of these agreements contain provisions relating to the environment. The WTO Secretariat boasts of them as the WTO's "green provisions" but does not define that term.[13]

In thinking about what renders a WTO provision green (i.e. pro-environmental), one should first recall the range of environmental policy instruments used by governments. They include: regulations, taxes, standards, labelling, subsidies and other technology incentives, trade controls, allocation and clarification of property rights, reporting and accountability for private actors, and environmental diplomacy. These instruments may be used to control pollution, manage natural resources or otherwise maintain the availability and quality of public goods.

Although WTO law does not directly dictate what the goals of a government's environmental policy should be or what instruments can be used, the scope of WTO law is broad enough to influence those choices in at least two ways. First, the WTO can influence environmental decision-making by facilitating economic growth through trade. The higher ensuing incomes *may* then lead to higher environmental quality by increasing the society's income and perhaps by catalysing greater public demand for environmental quality. Second, WTO law provides a background rule that removes policy space from governments to use environmental measures in certain ways. In other words, if environmental policy consists of active measures to achieve chosen environmental goals, then WTO law consists of passive restraints on the measures used. I suggest that a third mode of influence is also feasible: WTO law should be used to promote better environmental outcomes.

I shall now provide an overview of the environmental provisions present in WTO law, and some that are notably absent. WTO law contains three discrete areas of law, pertaining to trade in goods, trade in services, and trade-related intellectual property. The three areas are subsumed under the umbrella WTO treaty and share a common dispute settlement mechanism. I will discuss each area in turn.

4.1. Trade in goods

In applying its environmental policy to imported goods/products, a government must ordinarily follow the principles of most-favoured-nation (MFN) and national treatment. MFN treatment means that an imported product from a WTO member is not to be treated less favourably than a like imported product from any other country. National treatment means that an imported product is not to be treated less favourably with respect to a regulation than the like domestic product. With taxes, the rule is similar but a bit more strict. Although the WTO Secretariat has taken the position that regulations and taxes cannot be hinged on the upstream effects of production,[14] no authority exists in trade law for that proposition, and many environmentalists hope that WTO law will be flexible enough to accommodate such process-related measures. Another major trade rule for imported products is that quantitative restrictions such as import bans are generally prohibited. This rule would seem to apply to import bans dictated by a multilateral environmental agreement (MEA), but that point has not yet arisen in dispute settlement.

If a government has a good reason for violating MFN, national treatment or the prohibition of import (or export) bans, that government may be able to defend its measure by qualifying for one of the exceptions in GATT Article XX (GATT 1947). The WTO Secretariat sometimes forgets this.[15] Two exceptions are most applicable to environmental policy: Article XX(b) for measures "necessary to protect human, animal or plant life or health" and XX(g) for measures "relating to the conservation of exhaustible natural resources if such measures are made effective in conjunction with restrictions on domestic production or consumption". Both exceptions are subject to the requirement in the Article XX chapeau that "such measures are not applied in a manner which would constitute a means of arbitrary or unjustifiable discrimination between countries where the same conditions prevail, or a disguised restriction on international trade". Under WTO case law, a government seeking to claim one of these environmental exceptions has the burden of proof.

The trend in Article XX case law is for a more economic-based interpretation of the term "necessary". In the *Korea Beef* case, the Appellate Body held that, for a measure to be "necessary", it has to pass a "weighing and balancing process" in which a Panel in every case has to consider three "factors": (1) the importance of the value protected by that law or regulation, (2) the contribution made by the contested measure to the end pursued, and (3) the restrictive impact of the measure on imports.[16] Furthermore, the Appellate Body stated that this weighing and balancing process is comprehended in the determination of whether there is a WTO-consistent or less-WTO-inconsistent measure available that the government could reasonably be expected to employ.

This weighing and balancing test was not part of pre-1995 trade jurisprudence and has troubling implications for national health or environmental policy. One problem is that the test necessitates inter-country comparisons of utility in weighing, say, the health of one country versus the trade of another. Although national courts will sometimes weigh domestic health versus trade, having an international court do inter-country weighing is unusual. Because this task goes beyond what one would expect to be within the scope of a world trade court, the evolution of WTO case law may show that the Appellate Body is simultaneously also becoming a world court with jurisdiction over health and environment.

In addition to qualifying for a GATT General Exception, governments may derogate from the MFN requirement through three kinds of preferential trade arrangements: customs unions, free trade agreements (FTAs), and the generalized system of preferences (GSP) for developing countries. The establishment of customs unions has sometimes been accompanied by positive environmental harmonization, the leading example being the European Union and its predecessor communities. Some FTAs have included environmental cooperation, the leading example being the North American Free Trade Agreement and its parallel side agreement. The only GSP programme with an environmental component is the European Union's programme. Since 2001, it has included "special incentive arrangements for the protection of the environment", which apply to products of a tropical forest originating in countries that effectively apply national legislation that incorporates internationally acknowledged standards and guidelines concerning sustainable management of tropical forests (Council of the European Union 2001: Articles 21–24).

So far, this GSP environmental arrangement has not been challenged in WTO dispute settlement. Nevertheless, when India won its challenge in 2004 against the feature of the European GSP related to drug production and trafficking, the Appellate Body held that the WTO "enabling clause" for GSP requires a tariff-preference-granting country to "respond positively" to the particularized "development, financial and trade needs of developing countries".[17] This holding can be read as permitting the European Union's preference relative to products from sustainably managed tropical forests if sustainable timber management is considered to be a development need. If sustainable timber management is not considered a development need, then the Appellate Body's holding would seem to disallow that sort of environmental condition in a GSP programme.

In addition to being subject to the GATT, environmental regulations applying to imported products will also be subject to the WTO Agreement on Technical Barriers to Trade (TBT Agreement). This Agreement has numerous rules, only a few of which will be discussed here. A technical regulation is broadly defined as a government document laying down product characteristics or their "related processes and production methods" (TBT Agreement 1994: Annex 1.1). One core rule is that a governmental regulation "shall not be more trade-restrictive than necessary to fulfil a legitimate objective, taking account of the risks non-fulfilment would create" (TBT Agreement 1994: Article 2.2). Although no case law yet exists, one expert has argued that this rule requires that, when a regulation is claimed to be based on science, the regulator will need to have a risk assessment (see Motaal 2004: 857–859).

Another core TBT rule is that, when a relevant international standard exists, a government's technical regulation shall use that international standard as a basis for its regulation, unless the standard would be "an ineffective or inappropriate means" for the fulfilment of a legitimate objective (TBT Agreement 1994: Article 2.4). Standards are defined broadly and include environmental product standards. A "legitimate objective" is defined to include "protection of human health or safety, animal or plant life or health, or the environment" (TBT Agreement 1994: Article 2.2). Although a textual reading of the TBT Agreement suggests that its rules on international standards apply only to standards based on consensus, the Appellate Body has held that no consensus is required and thus that a standard determined through voting will be enforceable by the WTO.[18]

The commentary on the TBT Agreement emphasizes how the rule on international standards can undermine a government's effort to employ a regulation that seeks a higher level of protection than the international standard. Yet one should also recognize that this rule could possibly work in the opposite direction too. That is, the TBT Agreement could require laggard governments to move up to an international standard. Note, however, that the TBT Agreement (Article 12.4) states that developing countries may adopt regulations "aimed at preserving indigenous technology and production methods and processes compatible with their development needs" and that "developing country Members should not be expected to use international standards ... which are not appropriate to their development, financial and trade needs".

The TBT Agreement also contains rules to encourage governments to provide regulatory assistance to developing countries. Assistance is to be provided on the "preparation" of regulations and on the "methods" by which regulations can be met (TBT Agreement 1994: Articles 11.1, 11.3.2). So far, very little implementation has occurred.

For certain health-related regulations, TBT rules are supplanted by the Agreement on the Application of Sanitary and Phytosanitary Measures (SPS Agreement). Any measure covered by the SPS Agreement is carved out of the TBT Agreement. Given the considerable literature on SPS rules and their relation to biosafety and precaution,[19] and the new analyses elsewhere in this volume (see, e.g., Gupta, Chapter 2), the discussion here on SPS will be brief.

The SPS Agreement applies to regulations or import bans used to protect human, animal or plant life from a specific list of risks. A WTO member government may choose its desired level of protection, but "shall avoid arbitrary or unjustifiable distinctions in the levels it considers to be appropriate in different situations, if such distinctions result in discrimination or a disguised restriction on international trade" (SPS Agreement 1994: Article 5.5).[20] Member governments have considerably less autonomy in selecting SPS measures. Such measures are to be based on scientific principles and not maintained without sufficient scientific evidence, but the Agreement contains a clause (SPS Agreement 1994: Article 5.7) to provide flexibility in instances where relevant scientific evidence is insufficient. According to the Appellate Body, the precautionary principle "finds reflection" in that clause.[21]

The SPS Agreement privileges international standards set by the Codex Alimentarius Commission, by the International Office of Epizootics, and under the auspices of the International Plant Protection Convention. Governments must base their SPS measures on such standards, but may seek a higher (or lower) level of protection if there is a scientific justification or if the national standard meets all other SPS rules, including a trade-restrictiveness requirement that was drafted to be less onerous than the one in the TBT Agreement. Because the SPS Agreement relies upon international standards that need not be consensus standards (e.g. Codex standards), the WTO has the potential to become the enforcer of rules that not all WTO members have accepted.

Another policy instrument governed by the WTO is a government subsidy. The Agreement on Subsidies and Countervailing Measures (SCM Agreement) prohibits non-agricultural subsidies that have specificity and that cause "adverse effects to the interests" of WTO member countries (SCM Agreement 1994: Article 5). Originally, the SCM Agreement exempted certain environmental subsidies from this prohibition, but that derogation expired at the end of 1999, and WTO governments did not renew it (SCM Agreement 1994: Articles 8.2(c), 31). The exempted subsidies were for assistance to promote adaptation of existing facilities to new environmental requirements. According to the WTO Secretariat, the original provision was "intended to allow Members to capture positive environmental externalities when they arose".[22] Its expiration leaves subsidies to correct market failure subject to being challenged as WTO violations. So far, none has been.

Agricultural subsidies are governed by complex rules in the Agreement on Agriculture (1994), which commit countries to limit and reduce subsidies. For some environmental subsidies that have no trade-distorting effects, no reductions in support are required (the so-called "green box"). The Preamble of the Agreement on Agriculture suggests that its commitments have been made with regard to "the need to protect the environment".

4.2. Trade in services

The General Agreement on Trade in Services (GATS) can have significant environmental consequences. A key environmental plus is that the GATS may help enable governments to be more open to the importation of environmental services. The GATS also facilitates the movement of natural persons both to consume services (e.g. to attend a foreign university to study environmental science) and to deliver services (e.g. trained environmental technicians who provide assistance in another country).

Counterbalancing these positive repercussions from the GATS are the new disciplines that governments agree to accept. Environmental measures in the form of regulations, taxes or import bans will be subject to numerous GATS rules. For the most part, the GATS rules apply only to sectors where a government makes commitments.

In contrast to the GATT, which has two environment-related general exceptions, the GATS has only one. That exception applies to measures necessary to protect human, animal or plant life or health. This means that the GATT's environmental exception for conservation does not exist in the GATS (Waskow 2003: 793–795).

Although no environment disputes have yet arisen in the GATS, the absence of a conservation exception may make it hard to defend an environmental regulation subject to a dispute in the WTO. Challenging an environmental regulation was made easier by a recent Appellate Body decision holding that criminal laws prohibiting noxious services can be considered a zero quota that violates GATS Article XVI (Market Access).[23] This surprising holding came in the *Gambling* decision, in which three US laws banning Internet gambling were found to violate Article XVI, despite the fact that they were applicable *de jure* to domestic as well as to cross-border gambling services.

The governments that negotiated the GATS missed an opportunity to accord deference to the environment regime in the same way that deference is accorded to other regimes. For example, the GATS provides full deference to the rights and obligations of members of the International Monetary Fund and full deference to multilateral agreements to avoid double taxation (GATS 1994: Articles XI.2, XIV(e)). No analogous provisions exist for the environment.

The GATS does not define "services", an omission that has led observers to question whether certain environmental rights are to be considered services under the Agreement. For example, does a GATS-covered service include a right to pollute (e.g. an emission reduction unit), a right not to be polluted, or a right to exploit a natural resource (e.g. a fishery quota)? In my view, such government-created rights are not covered services, but no official interpretation yet exists.

4.3. Intellectual property

The third fount of substantive WTO law is the Agreement on Trade-Related Aspects of Intellectual Property Rights (TRIPS Agreement 1994). This Agreement mandates a minimum level of intellectual property protection that WTO member governments must provide to nationals of other WTO members. On some matters, TRIPS mandates that governments follow certain requirements of pre-existing intellectual property treaties. On other matters, TRIPS prescribes its own minimum requirements (UNCTAD–ICTSD 2005).

Patenting is the field of intellectual property most likely to have a significant effect on environmental and health quality. Under the TRIPS Agreement, governments are required to issue patents in all fields of technology, but "may exclude from patentability inventions, the prevention within their territory of the commercial exploitation of which is nec-

THE WTO AS AN ENVIRONMENTAL AGENCY   175

essary to protect *ordre public* or morality, including to protect human, animal or plant life or health or to avoid serious prejudice to the environment, provided that such exclusion is not made merely because the exploitation is prohibited by their law" (1994: Article 27.2).[24]

The effects of TRIPS on the environment will likely be mixed. A positive effect on the availability of technology is to be expected if longer patent terms lead to more innovation. On the other hand, a negative effect may ensue in lower-income countries if there are higher costs of obtaining products of foreign innovation (Nadal 2005: 22). The technology attracting almost all of the attention up until now has been pharmaceuticals (see Abbott 2005).

Despite its authority to cooperate with other international organizations, the Council for TRIPS has failed to act on some requests for observer status by major environmental agencies. For example, the Council has not given observer status to the United Nations Environment Programme or to the Secretariat for the Convention on Biological Diversity.

4.4. The WTO's structural provisions

Although the environment is not mentioned in the Agreement Establishing the WTO beyond the text of its Preamble, and is not mentioned in the WTO Dispute Settlement Understanding, there are two ways in which the WTO's structural provisions have implications for the environment.

For those countries that were not original members of the WTO, joining the WTO comes through an accession negotiation. A country seeking to join may do so only "on terms to be agreed between it and the WTO" (WTO 1994b: Article XII.1). This provision makes clear that it is the WTO itself that has the authority for proposing the entry terms. Because almost every country today wants to join the WTO (even North Korea has now sought observer status), the WTO has considerable leverage in those accession negotiations. Unfortunately, there is little public debate as to how that bargaining surplus should be used.

In the biggest accession negotiation so far, that of China, the WTO used its leverage to insist on both WTO-minus and WTO-plus provisions. WTO-minus provisions are when the WTO asks the applicant country to forgo certain rights that it would normally enjoy as a member. For example, the WTO did this on textiles and apparel trade in order to allow WTO members to engage in protectionist practices toward China for several years (Financial Times 2005: 18). WTO-plus provisions are when the WTO asks the applicant government to agree to rules beyond those required of WTO members. For example, the WTO did this to China in asking it to commit not to impose performance requirements of any kind on inward foreign investment (Qin 2003: 503).

Because of its position of power, the WTO can use its bargaining leverage for any issue it wants. Ideally, the WTO should use that leverage for a public benefit. In particular, the WTO should be promoting general interests rather than special interests. Telling general from special may not always be easy, but giving balm to European and US textile manufacturers hardly seems a general interest. Instead, the WTO should have used its leverage to promote political freedom in China. Another possibility was to use its leverage to convince China to remediate its industrial practices that cause harmful environmental effects on other countries (see Greenwire 2004).

The other WTO structural provision relevant here is the compliance measures available in the dispute settlement process if a WTO member loses a case and refuses to comply. Should that occur, the winning plaintiff may vindicate its victory by suspending trade concessions or other WTO obligations. For example, in the *Hormones* case, because the European Union has not complied, the United States and Canada are imposing 100 per cent tariffs against a range of goods. Yet, under current WTO practice, no review occurs of what products a government chooses to target. Perhaps a review should be undertaken of the projected environmental or human rights impact of the anticipated trade sanctions.

That completes the discussion of the most significant environmental features of the WTO treaty and the emerging case law. The other way in which the WTO has become an environmental agency is in including some environmental issues in the negotiating agenda for the Doha Round. Section 5 covers these developments.

5. The environment in WTO negotiations

Although previous multilateral trade rounds had given marginal consideration to the environment, the Doha Round marks the first time that the environment has been explicitly included on the negotiating agenda. To be sure, the environment is only a small aspect of the Round. Nevertheless, these features are important and will be discussed below.

5.1. Fishing

The Doha agenda commits governments to "clarify and improve WTO disciplines on fisheries subsidies, taking into account the importance of this sector to developing countries" (WTO 2001: para. 28). Although one could conceptualize a negotiation on fisheries subsidies as merely a commercial issue, the Doha Ministerial Declaration cross-lists the negotiations under the category of "Trade and environment". That makes sense because there are significant environmental benefits of removing subsidy-driven trade distortions in the fisheries sector (see WTO Secretariat 2000). Indeed, the 2005 WTO Annual Report characterizes the negotiations as being "aimed at restricting environmentally harmful fishing subsidies" (WTO 2005: 153). For some analysts, the fishery negotiations go too far in flirting with environmental conditionality (see Grynberg and Rochester 2005).

5.2. Environmental goods and services

Another important environment-related issue on the Doha agenda is the negotiation for the reduction or elimination of tariff and non-tariff barriers to environmental goods and services (Sampson 2005: 141). Although such negotiations are a trade liberalization objective, they are also an environmental objective, and the environmental benefit may be just as significant as the trade benefit. After all, current barriers to trade in, say, pollution control technology could not possibly be beneficial for the environment.

5.3. Win–win–win scenarios

A third environment-related feature is attention to "situations in which the elimination or reduction of trade restrictions and distortions would benefit trade, the environment and development" (WTO 2001: para. 32(i)). This provision was welcomed by environmentalists, who saw in it the possibility of WTO scrutiny of particular sectors and were pleased with the allusion to a "win–win–win" scenario (which in the business community is termed the "triple bottom line"). In a meeting of the Committee on Trade and Environment in special session in May 2007, the governments discussed a "non-paper" by a group of high-income countries identifying 153 environmental goods that could be negotiated. Unfortunately, the paper itself is being kept confidential by the WTO Secretariat.[25]

Although some sectoral policy was written into the WTO treaty – most notably in agriculture, textiles and clothing, and telecommunications – not much consideration has been given to reorganizing the WTO's environment work into sectors. Several sectors could benefit from more integrated attention, including, for example, aquaculture and fisheries, chemicals, energy goods and services, environmental goods and services, forestry, mining, tourism, and transport. For each sector, governments could consider how to improve environmental quality through WTO rules on subsidy reduction, regulations and standards on goods, regulations and standards on services, and technical assistance for developing countries. In that connection, the WTO could develop a list of recognized standard-setting bodies engaged in the development of environmental standards (see Chambers 2004: 81).

5.4. Multilateral environmental agreements

WTO members are negotiating on the relationship between WTO rules and specific trade obligations set out in multilateral environmental agreements (MEAs). This issue is important because, although MEAs have been using trade controls for over a century, there is a body of opinion inside the WTO that such controls are a violation of WTO rules and should no longer be permitted as environmental instruments. Many WTO member governments probably agree with Alan Oxley, a former GATT Council chairman, who has criticized leading MEAs for using "trade coercive measures" that disregard "national sovereignty" (Oxley 2004: 93–96). That opposition to trade measures in MEAs seems to have deterred the inclusion of trade controls in new MEAs. Other than the Stockholm Convention on Persistent Organic Pollutants (2001), no recent MEA contains specific trade obligations.

Although there was some hope by environmentalists that this threat to MEAs could be eliminated in the new trade round, the Doha agenda is highly circumscribed and is unlikely to lead to any fruitful outcome. Specifically, the governments have precluded any negotiation on trade measures applying to *non-parties* to the MEA and any result that would "add to or diminish the rights and obligations of members under existing WTO agreements" (WTO 2001: para. 32). In other words, the negotiators cannot propose changes to WTO rules.

When MEAs apply trade measures to non-parties, they can do so in two ways. One is to apply the same measure to a non-party as the MEA applies to a party (e.g. CITES). The other is to apply a discriminatory measure against a non-party (e.g. the Montreal Protocol on Substances that Deplete the Ozone Layer). Both are controversial within the WTO, but the second, involving discrimination, is more controversial. This stance seems hypocritical because the WTO provides space for discrimination against its non-parties. WTO member governments are permitted to discriminate against non-members with impunity. Even worse, when the WTO negotiates an accession agreement with a non-member (e.g. China), the WTO may insist that the applicant country accepts discrimination against it as a condition for joining.

Recently, a team of environmental analysts offered a good suggestion for "shifting the hapless debate within the CTE around MEAs toward a useful purpose" (Carpentier et al. 2005: 249). They recommend that the WTO look at each MEA and consider what particular trade liberalization, in goods and services, would help to meet the objective of that MEA.

5.5. TRIPS and biodiversity

Although not listed as an environmental issue, the relationship between the TRIPS Agreement and the Convention on Biological Diversity is included on the Doha agenda as an action item for the WTO Council on TRIPS. Specifically mentioned are the rules for patentability of plants and animals other than micro-organisms and for patentability of traditional knowledge and folklore. No decision has been reached by governments to commence negotiations.

5.6. Environmental reviews

The Doha Ministerial Declaration tasks the WTO Committee on Trade and Development and the Committee on Trade and Environment each to act, within their respective mandates, "as a forum to identify and debate developmental and environmental aspects of the negotiations" (WTO 2001: para. 51). Immediately after Doha, hopes were high in the environment community that this mandate would lead to a careful process of environmental impact assessment of proposed negotiating outcomes. Aaron Cosbey from the International Institute for Sustainable Development proposed several options for how the two WTO committees could carry out such efforts (Cosbey 2002). Unfortunately, neither committee initiated a robust assessment process. Doing so now would not be too late.

Back in 2002, the Johannesburg Plan of Implementation arising out of the World Summit on Sustainable Development called for "urgent action" to "support the successful completion of the work programme contained in the Doha Ministerial Declaration" (United Nations 2002: para. 47). The UN conference was correct to see the importance of successful WTO negotiations for the goal of sustainable development. Unfortunately, owing to various machinations at the WTO, the negotiators missed their 2004 deadline and the talks may continue to drag on for years.

6. Toward a new paradigm for the WTO

In this section I present a new paradigm for conceptualizing the WTO's role with respect to the environment. The existing paradigm is trade linkage, which considers how an organization with a trade purpose should deal with non-trade objectives, such as the environment. The new paradigm is to see the WTO as an organization with multiple objectives.

6.1. The multifunctional international organization

In a decision issued in 1996, the International Court of Justice decided, by a vote of 11 to 3, that it could not respond to a request by the World Health Organization (WHO) for an advisory opinion regarding the Legality of the Use by a State of Nuclear Weapons in Armed Conflict (International Court of Justice 1996).[26] The Court pointed to two reasons: the "general principle of speciality" and the logic of the overall system contemplated by the UN Charter (para. 26). On the same day that it turned down the WHO, the Court issued an advisory opinion on a similar question requested by the UN General Assembly.

With regard to the first reason for turning down the WHO, the Court held:

> International organizations are governed by the "principle of speciality", that is to say, they are invested by the States which create them with powers, the limits of which are a function of the common interests whose promotion those States entrust to them. (para. 25)

The Court further explained that, although the powers conferred on international organizations are normally the subject of an express statement in their constituent instruments, "the necessities of international life may point the need for organizations, in order to achieve their objectives, to possess subsidiary powers which are not expressly provided for in the basic instruments which govern their activities" (para. 25).

How does the international law principle of speciality relate to the environment? In my view, the environmental and market interdependence of life on Earth makes it hard to slice up distinct roles for environmental and economic agencies. Eventually, the bureaucratic preference for compartmentalization has to give way to environmental, economic and political realities.

6.2. Achieving an environmentally sound WTO

Consider the case of the WTO. Perhaps the governments drafting the WTO originally intended to create a trade-specific agency. Nevertheless, by the time the negotiations were completed in 1994, the Preamble to the WTO Agreement embraced sustainable development and the environment as a common interest. Then, in 1998, the Appellate Body breathed life into the Preamble language. In 2001, at the Doha Ministerial, the necessities of international life pointed to a need to launch new negotiations on trade and environment.

Maintaining a trade-only identity for the WTO was difficult because various non-trade issues, such as intellectual property, have already become part of the WTO's mission. Unlike intellectual property, however, where there existed a World Intellectual Property Organization fully competent in the field, for the environment there is no World Environment Organization with competence for major environmental issues (Speth 2004: 177). Thus, if the mandate of an international organization is driven by speciality and a rational division of labour, then the absence of a World Environment Organization provides more justification for a WTO role on environment than was justified for intellectual property.[27]

In calling the WTO an environmental agency, I am not suggesting that the principle of speciality has become obsolete. Even in today's interconnected world, many international agencies will remain highly specialized. What I am saying is that we should move beyond the constructs of the past that see the functional international organization as unitary in purpose. Instead, we should anticipate that major international organizations will often have multifunctional roles that may not always reflect full agreement among state members regarding the common interests that underlie the organization. Internal organizational complexity and divergence are to be expected (Coicaud 2001: 524–525). With member states each having multiple policy objectives, and with differing policy chromatograms for each state, it seems unreasonable to imagine that those same states will funnel down their differences to create single-function international agencies.

Visualizing the WTO as an environmental agency should become the new paradigm for integrating trade and environment. For many years, the operative paradigm has been "linkage" or "trade-and", with the trade regime being asked from the outside to give up some trade progress for the benefit of a different policy realm. Not surprisingly, the trade regime has often resisted the intrusions and congratulated itself for being so virtuous. Never mind that many of the governments inside the WTO have been tripping over themselves to hang on to as much protectionist trade policy as they can. As two WTO scholars recently remarked, "the reality ... is that the WTO is as much about protectionism as it is about free trade" (Guzman and Pauwelyn 2005: 7).

Staying with the old paradigm will frustrate a reconciliation of environment and trade objectives. Some who would resist seeing the WTO as an environmental agency might say that the WTO should maintain its singular trade mission but should improve its cooperation and coordination with environmental agencies. At best, that model seems to suggest that, when the WTO and, say, the Cartagena Biosafety Protocol are going in the same direction, they should hold hands and walk together. That is fine with me, but what I am really concerned about is what to do if the WTO and the Protocol go in different directions.

The prescription of cooperation – namely that the two organizations should work out their differences – is hard to operationalize when the purposes of the organizations differ. Therefore, we need a new consciousness at the WTO. The new consciousness should be that environment and sustainable development are part of the purpose of the WTO, not just a rhetorical adornment.

To make the WTO a better environmental performer, the mainstream environmental agencies, such as the United Nations Environment Programme, should seek to hold the WTO accountable as an environmental agency. These agencies should evaluate the WTO on its environmental achievements and its shortcomings. Furthermore, these agencies should work to internalize their environmental norms into WTO processes. Environment ministers should reflect on the fact that the trade community is not shy about insinuating its norms into environment treaties. This happened, for example, in the Cartagena Protocol (Oberthür and Gehring, Chapter 5 in this volume) and in the 1997 amendments to the International Plant Protection Convention (Article XVI).

One way that environmental agencies might help the WTO is by seeking to transplant their scientific orientation into the WTO. The WTO needs outside influence to convince it to make sure that all of its trade rules have a scientific basis. Take anti-dumping investigations for example. The WTO actually requires governments to perform such investigations (Anti-Dumping Agreement 1994: Article 5.1), and the WTO Secretariat has been generous in delivering technical assistance to developing countries to get their anti-dumping programmes into action. Yet there is no scientific basis for the notion that countries can boost their national income by imposing tariffs to stop the importation of low-price, "dumped" goods (see Irwin 2002: 124–128). To be sure, an anti-dumping programme can effectuate a redistributional objective within a country, but there are less trade-restrictive ways to accomplish that objective than blocking imports.

7. Conclusion

The WTO Secretariat contends that the WTO "is not an environmental protection agency" and that statement provides a good window into understanding how the WTO interacts with the environment. As this chapter has shown, the WTO is an environmental agency in some of its treaty provisions and in its pro-environment negotiating agenda. This agenda includes increasing market access for environmental goods and services and curtailing government subsidies that lead to over-fishing.

Why then does the Secretariat deny the WTO's environmental identity? It is because the WTO wants the power to tell governments what measures they cannot use for the environment, but wants to leave to national and international environmental agencies the responsibility for formulating strategies to address environmental problems and, on transborder threats, getting governments to agree. This may sound like a rational division of labour until one realizes that the WTO views its role as being constitutional on the international plane. What I mean by "constitutional" here is that whatever strategies emerge from environmental agencies are reviewable at the WTO.

In view of the disorganization and weak nature of international environmental governance, there is a danger in giving the WTO power over environmental measures without any responsibility for environmental outcomes. Reform can come through inculcating a greater sense of environmental responsibility in the WTO. By calling it an environmental agency we can challenge it to improve.

Today, the WTO operates as an environmental agency and yet is a poorly performing one: it allowed the Doha Round to languish despite the importance of trade liberalization for reducing world poverty; it made all environmental technology subsidies potentially actionable; it neglected to undertake environmental assessments of proposals in the Doha Round negotiations; its emerging case law threatens to reduce domestic environmental regulatory authority. Turning this around will not be easy. In *Spaceship Earth*, 40 years ago, Barbara Ward pondered reaching a time when we "realise the moral unity of our human experience and make it the basis of a patriotism for the world itself" (Ward 1966: 148). Attention to the world's ecological needs ought to be a hallmark of a world trade organization. Looking ahead a decade or two, one can hope that the WTO will not only become a better environmental agency but also be happy to admit it.

Acknowledgements

Thanks to Aaron Cosbey for his helpful comments.

Notes

1. Some of the mainstream studies and collections include: Anderson and Blackhurst (1992); Blackhurst et al. (1994); Esty (1994); Fredriksson (1999); Barrett et al. (2000); Könz (2000); Chambers (2001); Esty (2001); Figueres Olsen et al. (2001); Rao (2001);

Sampson and Chambers (2002); Steinberg (2002); Wallach (2002); Wiers (2002); Gaines (2003); Ishibashi (2003); Cosbey (2004a, 2004b); Knox (2004). The Principles elaborated in Blackhurst et al. (1994) were developed by a nine-person expert group that included Konrad von Moltke, as well as others, such as David Runnalls and Janine Ferretti, who were to make important contributions to the trade and environment field. Some recent studies include: Driesen (2005); Sampson and Whalley (2005); Zarrilli (2005).
2. With regard to trade flows, Copeland and Taylor (2003) argue that the scale, technique and compositional changes from trade can help to control pollution.
3. The proposition that the WTO is an environmental agency could be stated another way – namely, that certain WTO rules are part of international environmental law. Several years ago, a compendium of international environmental law, produced for Dutch universities, listed some trade law (see Lammers 1995: 235).
4. Some analysts argue that this extraneous role is a bad idea. See, e.g., Bhagwati (2004: 182–185).
5. But see the WTO Sutherland Commission report, which asserts that "[w]hile trade is an important factor in achieving development aims, the WTO is not a development agency" (Sutherland et al. 2005: para. 269).
6. The same point about trade supremacy can be made with respect to the human rights regime – there too the WTO has sometimes imagined itself as higher law (Pauwelyn 2003; Pruzin 2005).
7. Often this analysis has been directed at national environmental measures that seek to use trade access as a lever to change environmental policy in another country.
8. WTO Appellate Body Report, *United States – Import Prohibition of Certain Shrimp and Shrimp Products*, WT/DS58/AB/R, adopted 6 November 1998, para. 129 (emphasis in original, internal footnote deleted); see also paras 153, 155.
9. Ibid., para. 185 (emphasis in original).
10. WTO Panel Report, *United States – Import Prohibition of Certain Shrimp and Shrimp Products, Recourse to Article 21.5 of the DSU by Malaysia*, WT/DS58/R/W, adopted as modified by the Appellate Body, 21 November 2001, para. 5.54.
11. Calling the WTO a sustainable development agency may be one way to provide an overarching concept for the WTO's work on trade, environment and other functions. In correspondence with me, Aaron Cosbey (who was a close colleague of Konrad von Moltke for many years) makes that suggestion as a way "to take the logic of this paper to its final conclusion".
12. Appellate Body *Shrimp* Report, para. 154.
13. WTO, "The Environment: A Specific Concern", at ⟨http://www.wto.org/english/thewto_e/whatis_e/tif_e/bey2_e.htm⟩ (accessed 12 July 2007).
14. According to the WTO Secretariat, "trade restrictions cannot be imposed on a product purely because of the way it has been produced" (WTO, "The Environment: A Specific Concern").
15. According to the WTO Secretariat, "WTO Members are free to adopt national environmental protection policies provided that they do not discriminate between imported and domestically-produced like products (national treatment principle), or between like products imported from different trading partners (most-favoured-nation clause)" (WTO 2004: 7). This point is untrue because it ignores GATT Article XX, which may permit discrimination meeting the conditions in the Article XX chapeau.
16. Appellate Body Report, *Korea – Measures Affecting Imports of Fresh, Chilled and Frozen Beef*, WT/DS161,169/AB/R, adopted 10 January 2001, paras 163–166 (regarding Article XX(d)). The Appellate Body applied this test to health measures in the *Asbestos* case and has confirmed it twice since then, most recently in its April 2005 decision in the Dominican Republic *Cigarettes* case.

17. WTO Appellate Body Report, *European Communities – Conditions for the Granting of Tariff Preferences to Developing Countries*, WT/DS246/AB/R, adopted 20 April 2004, paras 162–165.
18. See TBT Agreement (1994: Annex 1.2 – explanatory note distinguishing between international standards and other standards covered by the Agreement); WTO Appellate Body Report, *European Communities – Trade Description of Sardines*, WT/DS231/AB/R, adopted 23 October 2002, para. 227.
19. For example, see Boisson de Chazournes and Thomas (2000); Charnovitz (2000); Covelli and Hohots (2003); Rivera-Torres (2003); Stewart and Johanson (2003); Motaal (2005).
20. This is the one discipline in the WTO that explicitly supervises the level of protection to be sought.
21. WTO Appellate Body Report, *European Communities – Measures Concerning Meat and Meat Products (Hormones)*, WT/DS26,48/AB/R, adopted 13 February 1998, para. 124. In a more recent case, the Panel noted that the Biosafety Protocol of 2000 had "confirmed the key function of the precautionary principle" in international law. WTO Panel Report, *Japan – Measures Affecting the Importation of Apples*, WT/DS245/R, adopted as modified by the Appellate Body, 10 December 2003, para. 5.34 & n. 161. This decision has been criticized for its strictness (see Vallely 2004).
22. WTO, "Agreement on Subsidies and Countervailing Measures", ⟨http://www.wto.org/English/tratop_e/envir_e/issu3_e.htm#scm⟩ (accessed 12 July 2007).
23. WTO Appellate Body Report, *United States – Measures Affecting the Cross-Border Supply of Gambling and Betting Services*, WT/DS285/AB/R, adopted 20 April 2005, paras 237, 238. A government could stay out of violation if it lists the national law in its negotiating schedule. This can shelter pre-existing environmental laws but would be useless for new environmental laws.
24. Exclusion is also possible for animals other than micro-organisms. See the complex rule laid out in TRIPS Article 27.3. See also the Convention on Biological Diversity, 5 June 1992, Article 16.5 (regarding intellectual property rights); text available at ⟨http://www.cbd.int/doc/legal/cbd-un-en.pdf⟩ (accessed 12 July 2007).
25. The confidential non-paper, circulated in document Job(07)/54, was entitled "Continued Work under Paragraph 31(iii) of the Doha Ministerial Declaration" (see WTO 2007: paras 159–160).
26. The Court's holding is criticized in the dissenting opinions and in scholarly commentary. For example, see the essays by Pierre Klein, Michael Bothe and Virginia Leary in Boisson de Chazournes and Sands (1999); see also the discussion in Burci and Vignes (2004: 114–118).
27. For a contrary view, see Maskus (2002).

REFERENCES

Abbott, Frederick M. (2005), "The WTO Medicines Decision: World Pharmaceutical Trade and the Protection of Public Health", *American Journal of International Law* 99(2): 317–358.

Agreement on Agriculture (1994), *Agreement on Agriculture. Annex 1A to the Final Act Embodying the Results of the Uruguay Round of Multilateral Trade Negotiations*, Marrakesh, 15 April 1994; available at ⟨http://www.wto.org/english/docs_e/legal_e/14-ag.pdf⟩ (accessed 12 July 2007).

Alvarez, José E. (2002), "The WTO as Linkage Machine", *American Journal of International Law* 96(1): 146–158.
Anderson, Kym and Richard Blackhurst, eds (1992), *The Greening of World Trade Issues*. Hemel Hempstead: Harvester Wheatsheaf.
Anti-Dumping Agreement (1994), *Agreement on Implementation of Article VI of the General Agreement on Tariffs and Trade 1994. Annex 1A to the Final Act Embodying the Results of the Uruguay Round of Multilateral Trade Negotiations*, Marrakesh, 15 April 1994; available at ⟨http://www.wto.org/english/docs_e/legal_e/19-adp.pdf⟩ (accessed 12 July 2007).
Barrett, Scott et al. (2000), "Special Issue: Trade and Environment", *Environment and Development Economics* 5(4): 341–530.
Bhagwati, Jagdish (2004), *In Defense of Globalization*. Oxford: Oxford University Press.
Blackhurst, Richard et al. (1994), *Trade and Sustainable Development Principles*. Winnipeg. International Institute for Sustainable Development.
Boardman, Robert (1981), *International Organization and the Conservation of Nature*. Bloomington: Indiana University Press.
Boisson de Chazournes, Laurence and Philippe Sands, eds (1999), *International Law, the International Court of Justice and Nuclear Weapons*. Cambridge: Cambridge University Press.
Boisson de Chazournes, Laurence and Urs P. Thomas (2000), "The Biosafety Protocol: Regulatory Innovation and Emerging Trends", *Swiss Review of International and European Law* 4: 513–557.
Burci, Gian Luca and Claude-Henri Vignes (2004), *World Health Organisation*. The Hague: Kluwer Law International.
Carpentier, Chantal Line, Kevin P. Gallagher and Scott Vaughan (2005), "Environmental Goods and Services in the World Trade Organisation", *Journal of Environment & Development* 14(2): 225–251.
Chambers, W. Bradnee, ed. (2001), *Inter-linkages. The Kyoto Protocol and the International Trade and Investment Regimes*. Tokyo: United Nations University Press.
Chambers, Bradnee (2004), "WTO and Sustainable Development", in Tatsuro Kunugi (ed.), *Taking Leadership in Global Governance*. Osawa: International Christian University, pp. 79–81.
Charnovitz, Steve (2000), "The Supervision of Health and Biosafety Regulation by World Trade Rules", *Tulane Environmental Law Journal* 13(2): 217–302.
Coicaud, Jean-Marc (2001), "International Organisations, the Evolution of International Politics, and Legitimacy", in Jean-Marc Coicaud and Veijo Heiskanen (eds), *The Legitimacy of International Organisations*. Tokyo: United Nations University Press, pp. 519–552.
Copeland, Brian R. and M. Scott Taylor (2003), *Trade and the Environment. Theory and Evidence*. Princeton, NJ: Princeton University Press.
Cosbey, Aaron (2002), "Taking the Doha Language Seriously: The WTO as if Sustainable Development Really Mattered", address prepared for the Royal Institute of International Affairs conference Sustainable Development in the New Trade Round: Trade, Investment and Environment after Doha, Chatham

House, May; available at ⟨http://www.iisd.org/pdf/2002/trade_riia_paper_may2002.pdf⟩ (accessed 12 July 2007).

Cosbey, Aaron (2004a), *Lessons Learned on Trade and Sustainable Development.* Winnipeg: International Institute for Sustainable Development.

Cosbey, Aaron (2004b), *A Capabilities Approach to Trade and Sustainable Development. Using Sen's Conception of Development to Re-Examine the Debates.* Winnipeg: International Institute of Sustainable Development.

Council of the European Union (2001), *Council Regulation (EC) No 2501/2001: Applying a Scheme of Generalised Tariff Preferences*, 10 December, *Official Journal of the European Communities*, L346; available at ⟨http://trade.ec.europa.eu/doclib/docs/2003/may/tradoc_113021.pdf⟩ (accessed 12 July 2007).

Covelli, Nick and Viktor Hohots (2003), "The Health Regulation of Biotech Foods under the WTO Agreements", *Journal of International Economic Law* 6(4): 773–795.

Customs Simplification Convention (1923), *Convention Relating to the Simplification of Custom Formalities*, 3 November 1923, 30 League of Nations Treaty Series 371.

Driesen, David M. (2005), "What Is Free Trade? The Rorschach Test at the Heart of the Trade and Environment Debate", in E. Kwan Choi and James C. Hartigan (eds), *Handbook of International Trade*, vol. 2. Malden, MA: Blackwell Publishing, pp. 5–41.

Esty, Daniel C. (1994), *Greening the GATT*. Washington, DC: Institute for International Economics.

Esty, Daniel C. (2001), "Bridging the Trade–Environment Divide", *Journal of Economic Perspectives* 15(3): 113–130.

Figueres Olsen, José María et al. (2001), "Trade and Environment at the World Trade Organization: The Need for a Constructive Dialogue", in Gary P. Sampson (ed.), *The Role of the World Trade Organization in Global Governance.* Tokyo: United Nations University Press.

Financial Times (2005), "Textiles Stitch-up: Whatever the EU and China Say, Their Deal Mocks Free Trade" (editorial), *Financial Times*, 14 June, p. 18.

Fredriksson, Per G., ed. (1999), *Trade, Global Policy and the Environment.* Washington, DC: World Bank.

Gaines, Sanford E. (2003), "The Problem of Enforcing Environmental Norms in the WTO and What to Do about It", *Hastings International & Comparative Law Review* 26: 321–385.

GATS (1994), *General Agreement on Trade in Services. Annex 1B to the Final Act Embodying the Results of the Uruguay Round of Multilateral Trade Negotiations*, Marrakesh, 15 April 1994; text available at ⟨http://www.wto.org/english/docs_e/legal_e/26-gats.pdf⟩ (accessed 12 July 2007).

GATT (1947), "Article XX: General Exceptions", *The General Agreement on Tariffs and Trade*; text available at ⟨http://www.wto.org/english/docs_e/legal_e/gatt47_02_e.htm#articleXX⟩ (accessed 12 July 2007).

Greenwire (2004), "China's Mercury Pollution Affects Entire Globe, Scientists Say", 17 December.

Grynberg, Roman and Natallie Rochester (2005), "The Emerging Architecture of a World Trade Organisation Fisheries Subsidies Agreement and the Interests of Developing Coastal States", *Journal of World Trade* 39(3): 503–526.

Guzman, Andrew and Joost Pauwelyn (2005), "An Insider's Guide to the WTO's Problems", *Bridges* 9(1): 7.

Havana Charter (1948), *Charter for an International Trade Organization*; text available at ⟨http://www.wto.org/english/docs_e/legal_e/prewto_legal_e.htm⟩ (accessed 12 July 2007).

Hobbs, Anna L., Jill E. Hobbs and William A. Kerr (2005), "The Biosafety Protocol: Multilateral Agreement on Protecting the Environment or Protectionist Club?", *Journal of World Trade* 39(2): 281–300.

International Court of Justice (1996), "Legality of the Use by a State of Nuclear Weapons in Armed Conflicts: Advisory Opinion of 8 July", available at ⟨http://www.icj-cij.org/docket/files/93/7407.pdf⟩ (accessed 12 July 2007).

Irwin, Douglas A. (2002), *Free Trade Under Fire*. Princeton, NJ: Princeton University Press.

Ishibashi, Kanami (2003), "Environmental Measures Restricting the Waste Trade", in Alexandre Kiss et al. (eds), *Economic Globalization and Compliance with International Environmental Agreements*. The Hague: Kluwer Law International.

Jackson, John H. (2005), "Justice Feliciano and the WTO Environmental Cases: Laying the Foundations of a 'Constitutional Jurisprudence' with Implications for Developing Countries", in Steve Charnovitz, Debra P. Steger and Peter van den Bossche (eds), *Law in the Service of Human Dignity*. Cambridge: Cambridge University Press.

Knox, John H. (2004), "The Judicial Resolution of Conflicts between Trade and the Environment", *Harvard Environmental Law Review* 28: 1–78.

Könz, Peider, ed. (2000), *Trade, Environment and Sustainable Development: Views from Sub-Saharan Africa and Latin America. A Reader*. Tokyo: United Nations University Press.

Krueger, Anne O., ed. (1998), *The WTO as an International Organization*. Chicago: University of Chicago Press.

Lammers, J. G. (1995), *Internationaal Milieurecht*. The Hague: T. M. C. Asser Instituut.

Maskus, Keith E. (2002), "Regulatory Standards in the WTO: Comparing Intellectual Property Rights with Competition Policy, Environmental Protection, and Core Labor Standards", *World Trade Review* 1(2): 135–152.

Motaal, Doaa Abdel (2004), "The 'Multilateral Scientific Consensus' and the World Trade Organisation", *Journal of World Trade* 38(5): 855–876.

Motaal, Doaa Abdel (2005), "Is the World Trade Organisation Anti-Precaution?", *Journal of World Trade* 39(3): 483–501.

Nadal, Alejandro (2005), "Redesigning the Trading System for Sustainable Development", *Bridges* 9(5): 21–22.

Oxley, Alan (2004), "The Relationship between MEAs and WTO Rules", in UNCTAD, *Trade and Environment Review 2003*. Geneva: UNCTAD.

Parliamentary Conference on the WTO (2003), "Final Declaration", Geneva, 18 February; available at ⟨http://www.ipu.org/splz-e/trade03.htm⟩ (accessed 12 July 2007).

Pauwelyn, Joost (2003), "WTO Compassion or Superiority Complex? What to Make of the WTO Waiver for 'Conflict Diamonds'", *Michigan Journal of International Law* 24: 1177–1207.

Pruzin, Daniel (2005), "U.N. Human Rights Official Warns against WTO Restrictions on Food Aid", *BNA Daily Report for Executives*, 20 July.

Qin, Julia Ya (2003), "'WTO-Plus' Obligations and Their Implications for the World Trade Organisation Legal System", *Journal of World Trade* 37(3): 484–522.

Rao, P. K. (2001), *Environmental Trade Disputes at the WTO*. Lawrenceville, NJ: Pinninti Publishers.

Rivera-Torres, Olivette (2003), "The Biosafety Protocol and the WTO", *Boston College International and Comparative Law Review* 26: 263–323.

Sampson, Gary P. (2005), "The World Trade Organization and Global Environmental Governance", in W. Bradnee Chambers and Jessica F. Green (eds), *Reforming International Environmental Governance*. Tokyo: United Nations University Press, pp. 93–149.

Sampson, Gary P. and W. Bradnee Chambers, eds (2002), *Trade, Environment, and the Millennium*, 2nd edn. Tokyo: United Nations University Press.

Sampson, Gary and John Whalley, eds (2005), *The WTO, Trade and the Environment*. Cheltenham, UK: Edward Elgar.

SCM Agreement (1994), *Agreement on Subsidies and Countervailing Measures. Annex 1A to the Final Act Embodying the Results of the Uruguay Round of Multilateral Trade Negotiations*, Marrakesh, 15 April 1994; available at ⟨http://www.wto.org/english/docs_e/legal_e/24-scm.pdf⟩ (accessed 12 July 2007).

Shaffer, Gregory C. (2001), "The World Trade Organization under Challenge: Democracy and the Law and Politics of the WTO's Treatment of Trade and Environment Matters", *Harvard Environmental Law Review* 25(1): 1–93.

Speth, James Gustave (2004), *Red Sky at Morning*. New Haven, CT: Yale University Press.

SPS Agreement (1994), *Agreement on the Application of Sanitary and Phytosanitary Measures. Annex 1A to the Final Act Embodying the Results of the Uruguay Round of Multilateral Trade Negotiations*, Marrakesh, 15 April 1994; available at ⟨http://www.wto.org/English/docs_e/legal_e/15-sps.pdf⟩ (accessed 5 July 2007).

Staiger, Robert W. (2004), "Report on the International Trade Regime for the International Task Force on Global Public Goods", February, ⟨http://www.gpgtaskforce.org/bazment.aspx?page_id=175⟩ (accessed 12 July 2007).

Steinberg, Richard H. (2002), *The Greening of World Trade Law*. Lanham, MD: Rowman & Littlefield.

Stewart, Terence P. and David S. Johanson (2003), "A Nexus of Trade and the Environment: The Relationship between the Cartagena Protocol on Biosafety and the SPS Agreement of the World Trade Organization", *Colorado Journal of International Environmental Law and Policy* 14: 1–52.

Sutherland, Peter et al. (2005), *The Future of the WTO. Report by the Consultative Board to the Director-General Supachai Panitchpakdi*. Geneva: WTO.

TBT Agreement (1994), *Agreement on Technical Barriers to Trade. Annex 1A to the Final Act Embodying the Results of the Uruguay Round of Multilateral Trade Negotiations*, Marrakesh, 15 April 1994; text available at ⟨http://www.wto.org/english/docs_e/legal_e/17-tbt.pdf⟩ (accessed 5 July 2007).

Trade Prohibitions Convention (1927), *Convention for the Abolition of Import and Export Prohibitions and Restrictions*, 8 November 1927, 97 League of Nations Treaty Series 391.

TRIPS Agreement (1994), *Agreement on Trade-Related Aspects of Intellectual Property Rights. Annex 1C to the Final Act Embodying the Results of the Uruguay Round of Multilateral Trade Negotiations*, Marrakesh, 15 April 1994; text available at ⟨http://www.wto.org/english/docs_e/legal_e/27-trips.pdf⟩ (accessed 12 July 2007).

UNCED (1992), *Agenda 21*; text available at ⟨http://www.un.org/esa/sustdev/documents/agenda21/english/agenda21toc.htm⟩ (accessed 12 July 2007).

UNCTAD–ICTSD (2005), *Resource Book on TRIPS and Development*. Cambridge: Cambridge University Press.

United Nations (2002), *Plan of Implementation of the World Summit on Sustainable Development* (Johannesburg Plan of Implementation); available at ⟨http://www.un.org/esa/sustdev/documents/WSSD_POI_PD/English/WSSD_PlanImpl.pdf⟩ (accessed 12 July 2007).

Vallely, Patrick J. (2004), "Tensions between the Cartagena Protocol and the WTO: The Significance of Recent WTO Developments in an Ongoing Debate", *Chicago Journal of International Law* 5: 369–378.

Von Moltke, Konrad (1993), "A European Perspective on Trade and the Environment", in Durwood Zaelke et al. (eds), *Trade and the Environment*. Washington, DC: Island Press, pp. 93–108.

Von Moltke, Konrad (1996), "The World Trade Organisation and the Environment: What Must Change", PSIO Occasional Paper, Graduate Institute of International Studies, Geneva.

Von Moltke, Konrad (2005), "Clustering International Environmental Agreements as an Alternative to World Environment Organisation", in Frank Biermann and Stephen Bauer (eds), *A World Environment Organisation. Solution or Threat for Effective International Environmental Governance*. Aldershot, UK: Ashgate, pp. 175–204.

Wallach, Lori M. (2002), "Accountable Governance in the Era of Globalization: The WTO, NAFTA, and International Harmonization of Standards", *University of Kansas Law Review* 50(4): 823–865.

Ward, Barbara (1966), *Spaceship Earth*. New York: Columbia University Press.

Waskow, David (2003), "Environmental Services Liberalisation: A Win–Win or Something Else Entirely?", *International Lawyer* 37(3): 777–799.

Wiers, Jochem (2002), *Trade and Environment in the EC and the WTO*. Groningen, The Netherlands: Europa Law Publishing.

WTO [World Trade Organization] (1994a), "Decision on Trade and Environment", adopted by ministers at the meeting of the Uruguay Round Trade

Negotiations Committee, Marrakesh, 14 April; text available at ⟨http://www.wto.org/English/docs_e/legal_e/56-dtenv.pdf⟩ (accessed 12 July 2007).

WTO (1994b), *Agreement Establishing the World Trade Organization*, Marrakesh; text available at ⟨http://www.wto.org/english/docs_e/legal_e/04-wto.pdf⟩ (accessed 12 July 2007).

WTO (2001), *Ministerial Declaration*, Ministerial Conference, Fourth Session, Doha, 9–14 November, WT/MIN(01)DEC/1, 20 November; available at ⟨http://www.wto.org/English/thewto_e/minist_e/min01_e/mindecl_e.pdf⟩ (accessed 12 July 2007).

WTO (2004), *Trade and Environment at the WTO*. Geneva: WTO; text available at ⟨http://www.wto.org/English/tratop_e/envir_e/envir_wto2004_e.pdf⟩ (accessed 12 July 2007).

WTO (2005), *Annual Report*. Geneva: WTO.

WTO (2007), *Summary Report on the Eighteenth Meeting of the Committee on Trade and Environment in Special Session, 3–4 May 2007*, Restricted, TN/TE/R/18, 8 June.

WTO Secretariat (2000), "Environmental Benefits of Removing Trade Restrictions and Distortions: The Fisheries Sector", WT/CTE/W/167, 16 October.

Zarrilli, Simonetta (2005), "International Trade in GMOs and GM Products: National and Multilateral Legal Frameworks", Policy Issues in International Trade and Commodities Study Series No. 29. Geneva: UNCTAD.

Zhang, Ruosi (2004), "Food Security: Food Trade Regime and Food Aid Regime", *Journal of International Economic Law* 7(3): 565–584.

8
Additional tributes to Konrad von Moltke

Adil Najam
Professor, The Fletcher School of Law and Diplomacy, Tufts University
Konrad von Moltke was truly – and not just literally – a giant in the field of international environmental politics. Indeed, he was amongst its pioneers. The journal *International Environmental Affairs*, which he edited for many years, not only was an early stomping ground for many of us but was instrumental in giving the field academic recognition and forcing an intellectual rigor on the study of global environmental politics. It played – and, I would argue, Konrad played (along with other pioneers such as Oran Young) – an absolutely critical role in making the field respectable for us younger and less adventurous researchers to venture into. He was an ultimate mentor to younger academics: kind, insightful, but always demanding of rigor and never "easy" on anyone. Over the years, I was fortunate in working closely with him on a variety of initiatives, particularly those related to trade and environment and on global environmental governance. Over a decade of interactions and innumerable meetings, I cannot remember too many occasions when I did not leave, saying to myself, "Why did I not think of that!"

Konrad was also a role model for all of us who wish to link scholarship with practice. His imprint on the practice of international environmental politics is quite profound. I remember having dinner with him, Klaus Töpfer (UNEP's former Executive Director) and some others at a meeting and Klaus saying something to the effect that Konrad was the "practice world's scholar of choice". Although made casually at a conference

---

Institutional interplay: Biosafety and trade, Young, Chambers, Kim and ten Have (eds), United Nations University Press, 2008, ISBN 978-92-808-1148-3

dinner, it was a very apt description. In many ways he was also the scholarly world's practitioner of choice. He was a wonderful, and vital, bridge between the two worlds – insisting on scholarly rigor in the pursuit of practice insights and on practical implications in the crafting of the scholarly agenda.

****

Dan C. Esty
Professor, Yale University
Konrad von Moltke was a true global citizen who understood the inescapable linkages that unite all people on the Earth. He cared deeply about finding successful strategies for addressing transboundary pollution and natural resource management challenges. He recognized that successful efforts had to work across the environment/economy divide and ensure both a more prosperous world and one that better protected Nature. Konrad von Moltke's towering presence in the realm of global environmental governance will be a source of inspiration for years to come.

****

Ernst von Weizsaecker
Dean, Donald Bren School of Environmental Science & Management, University of California, Santa Barbara
Konrad von Moltke had a fine sense of what is internationally important. In the 1970s he discovered the significance of European environmental policy and created an institute, the Institute for European Environmental Policy, to pursue this task. Later, he was one of the pioneers investigating conflicts between free trade regimes and environmental protection. Among these conflicts, one of the most exciting is surely the one on biosafety. Will the Cartagena Protocol survive the massive attacks launched by the biotech industry, which is using benevolent free traders as door openers for its business? Again, Konrad von Moltke was there. I would have been keenly interested to read his analysis!

****

Mark Halle
European Representative and Director, Trade and Investment, International Institute for Sustainable Development
Heroes are people we not only admire, but in many ways also use as models. In that sense, Konrad was a hero. I not only admired him, but studiously sought to imitate his fine balance between professional rigour and outward casualness; his propensity to surprise and delight. "Thinking out of the box" has become a cliché, but surprisingly few people are really able to do it. Konrad did it as a way of life, as a game, as a source

of fun. It was easy to look up to Konrad, both physically of course but also in terms of what he was. For someone as wise and worldly, his lack of interest in honours, in recognition, and in the normal trappings of success was sobering. His patrician background and his cosmopolitan sophistication might have predisposed him to a more classical academic itinerary. Instead, he sought one thing only in his professional life – the luxury to think and to work on the issues that he found fascinating. No title, no swollen income, no guarantee of comfort could replace this priority, at least in his professional life.

Konrad's departure leaves a big hole in our midst. We will never again have the easy benefit of his genius, or the thrill of his iconoclastic perspective, or the companionship of many travels together. I miss everything about him – his facial expressions, his quirks of speech ("I always say …"), his nervous spit curl, his broad back disappearing down the hall, pulling the ubiquitous suitcase on wheels. My life and my work have been immeasurably enriched by Konrad's passage through them.

****

Nigel Haigh
Former Director of IEEP London
On becoming the founder director of the Institute for European Environmental Policy (IEEP) in Bonn in 1976 Konrad focused on the institutional arrangements for environmental protection. He knew that proposing the right policies was not enough and that if the then rather weak European Parliament was to be a key player its powers had to be developed. His advocacy of amendments to the Treaty of Rome was just one of his contributions, and he had the satisfaction of seeing his ideas adopted. I was one of many people whose lives were changed by Konrad, in my case when he asked me to open the London office of IEEP in 1980. One of his many ideas that I recall was the analogy of billiard balls. Policy was often made, he said, by a proposal developed by one international institution ricocheting off another, and possibly another, until eventually it becomes reality. I was able to develop this idea for the case of acid rain in an article in the first issue of *International Environmental Affairs* – which Konrad edited. This showed – and I could never have done this without Konrad's impetus – that at least six separate international institutions were involved between 1970 and 1988.

****

Oran R. Young
Professor, Donald Bren School of Environmental Science & Management, University of California, Santa Barbara

Konrad's death – following the death several years back of Dana Meadows – is an occurrence that is somehow incomprehensible. The three of us, born in the same year, shared many things during our years at Dartmouth College. Not only were we committed to bringing scientific knowledge to bear on policymaking regarding large-scale environmental issues; we were also prepared cheerfully to take drastic steps, such as resigning tenured faculty positions, to gain the freedom needed to pursue this goal vigorously. We lived by our wits, an exhilarating albeit occasionally anxiety-producing situation that made it imperative to stay on the cutting edge far beyond the halls of academia.

The high point came during a period of years in the 1990s when we were able to take over some space in an old science building to create a vital center of international environmental affairs. Konrad was making seminal contributions to the environment and trade debate and editing the journal *International Environmental Affairs*. Dana was engaged in pioneering work on the idea of sustainable development. And I was in the thick of efforts to promote international cooperation in the Circumpolar Arctic, as well as working out the analytic foundations of the study of governance in world affairs. The result was magical. We stimulated each other's thinking, and shared ideas about communicating our ideas to a broader audience. And it worked. The evidence of our collective influence on governance for sustainable development in the twenty-first century is apparent to all who are familiar with this field. Not a day goes by that I do not think of Konrad and of our glory days at Dartmouth in pushing the envelope of the science/policy interface relating to governance for sustainable development.

****

Owen Cylke
Senior Policy Officer, WWF Macroeconomics Programme, Washington, DC

It is not often one finds a mentor at the age of 66, but I did – the relationship with Konrad emerging from a casual conversation in the ocean waves off Cancun during the fifth WTO ministerial conference in 2003. Konrad and I talked about trade, trade and environment, poverty and environment, the meaning of development, the state of the world, the role of our work in that world, the prospects for change, pathways of change, and our personal and professional histories. From that I came to know him as adviser, authority, backseat driver, coach, confidant, consultant, counsel, docent, doctor, dominie, don, educationist, educator, elder, expert, fellow, forerunner, great soul, guide, illuminator, instructor, intellect, intellectual, kibitzer, lover of wisdom, maestro, mahatma, man of intellect, man of wisdom, mandarin, mastermind, meddler, monitor, nestor, oracle, orienter, pandit, pathfinder, pedagogist, pedagogue, philosopher, preceptor, preparationist, preparator, preparer, professor, pundit, sage, sapient, savant, scholar, schoolmaster, seer, teacher, thinker, trailblazer, trainer, very wise man – and hopefully friend.

Just before he discovered his adversity earlier this year, we were scheduled to meet in Kenya to consider the significance of the flower trade for all of the questions we canvassed in Cancun. Sadly we were unable to meet, but the questions remain embedded in my (our) work and in the lives and thinking of those whom that work touches. Thank goodness for Konrad.

****

R. Andreas Kraemer
Director, Ecologic – Institute for International and European Environmental Policy

As a pioneering thinker, Konrad was an inspiration to the Ecologic. He took a particular interest in the development of international trade policy and law and other areas of the global economic order, applying the same principles that had served so well in the context of the EU. He also maintained his focus on networking civil society organizations and succeeded in linking the many institutions, including in academia, in which he had a role. Combining the rigour of the mathematician in him with the sense of proportion gained as a historian, he tirelessly worked to improve education on both sides of the Atlantic.

Konrad abhorred violence and did not seek conflict, but he did not shy away from political debate and was a formidable and intrepid discussant, with his views grounded not only in careful analysis but also in high moral and ethical principles. More than two metres tall and a founder and inspiration to many transnational academic and civil society networks, he frequently, and only half jokingly, referred to himself as the "largest multinational in the room". His lasting legacy is a multitude of networked bodies that make up part of a global civil society for sustainable development, peaceful conflict resolution and democracy based on grass-roots activism and involvement. His example and principles will continue to be a moral compass for our advocacy.

****

Richard G. Tarasofsky
Programme Head, Energy, Environment and Development Programme, Chatham House, London

Konrad von Moltke's contribution to international environmental policy cannot be overestimated. Not only was he an influential thinker and writer, but his involvement in the formative stages of important bodies, such as the International Institute for Sustainable Development (IISD), the Institute for European Environmental Policy (IEEP), and others, ensured that his imprint on the policy community was truly profound. Amongst Konrad's great insights was the recognition, very early on, that the fortunes of environmental policy are intrinsically linked to economic policy and economic institutions – i.e. that environmental policy would be both limited and boosted by how its means and objectives coincided with the economic agenda. This was the basis of his pioneering work on European Community environmental policy and on the interface between GATT/WTO and the environment – later widened to include development and investment. But, while pointing out the risks posed by other interests and agendas, he was also environmental policy's great champion. A key message throughout his writings and speeches was for the environmental community not to despair at the gravity of the obstacles; but on the contrary to celebrate its tremendous achievements, as well as the robustness and strength of the international environmental regime. This message needs to be constantly reiterated, and Konrad was tremendous in doing so.

Thus, it is very fitting indeed for a set of tributes to Konrad to appear in a book that deals with the biosafety regime. After all, the Biosafety Protocol is not only an important international success in using trade measures to achieve equitable environmental and developmental outcomes, but a triumph over powerful countries and industries that attempted to prevent it from coming into being.

****

Steve Charnovitz
Associate Professor, George Washington University Law School

Konrad von Moltke was unforgettable in so many positive ways. As a scholar, he was among the first to explore new issues, such as the trade and environment linkage. As a teacher, he displayed modesty and generosity, and shared his encyclopaedic mind. When he spoke as a panellist or workshop participant, everyone in the room would tune in for the big ideas and little witticisms to follow. He was interdisciplinary in method, internationalist in perspective, and passionately interested in how he could help people and the planet.

****

Thierry Lavoux
Former Director, Institute for European Environmental Policy (IEEP), Paris Office

For all those meeting Konrad for the first time, he was an imposing and intimidating figure. Immense in size, one was also struck by his beautiful face touched with nobility and virility. Konrad wanted the IEEP to be a forum close to the European institutions and the member states. At a time when debates about environmental taxation were stirring up opinion among experts in Western capitals, it should be remembered that at the very start of the1980s Konrad had launched the subject by suggesting that environmental policy could become more effective through economic tools; that, in this manner, environmental policy would enter the "court of great policies". Similarly, he felt strongly that the implementation of Community legislation in the member states was neglected, if not ignored. He succeeded, not without difficulties, in convincing European officials that his Institute, thanks to its offices located across Europe, could investigate the way in which the main Directives were implemented in the various national laws. This is the pioneering topic that quickly made the reputation of the IEEP. There was no need for him to speak louder to make himself understood, although I did see him getting irritated by some national or European officials who did not understand or did not share his views! Then Konrad left the Institute he had created – too soon in my opinion. He certainly died appallingly early, deeply saddening and disconcerting all those who had known him. He was "un grand Monsieur", as we say in French.

Index

"f" refers to figure; "n" to notes; "t" to table.

advance informed agreement (AIA), 11, 22, 26–29, 61, 66, 78, 108–9, 144
Agreement on Agriculture, 173, 185
Agreement on Subsidies and Countervailing Measures (SCM Agreement), 173, 189
Agreement on the Application of Sanitary and Phytosanitary Measures (SPS). *See* SPS Agreement
Agreement on the Conservation of Polar Bears, 55
AIA. *See* advance informed agreement (AIA)
Andersen, Regine, 7
Antarctic Treaty System, 55
Anti-Dumping Agreement, 182
Argentina, 16, 22, 27, 37–38, 41n3, 63–64, 108
Australia, 16, 27, 41n3, 42n6, 63–64, 108

Basel Convention on hazardous wastes, 21, 41n2, 42, 62, 86, 97–98
"beef hormone" dispute, 21, 25–26, 36, 41n1, 85
biodiversity. *See also* genetically modified organisms (GMOs); "living modified organisms" (LMOs)
  attempts to preserve, 153

biotechnology, long term effects, 147
Cartagena Protocol on Biosafety, 30, 31t2.1, 64, 65t3.3, 82, 87, 90n7, 90n13, 91, 109, 112, 141, 145, 163, xi
climate regimes and horizontal interplay, 53
Convention on Biological Diversity (CBD), 26–27, 50, 71, 78, 94, 132, 141, 146, 153–54, 175, 179, 185n24
ecosystems and loss of, 49, 151
General Agreement on Tariffs and Trade (GATT), 105
GMOs, risk assessment, 82
horizontal interplay and climate regimes, 64, 82
international trade rules on, 163
public health and, 139
scientific uncertainty, 23, 29–30, 59–62, 80–81, 83, 89, 90n7, 90n12, 141, 149
SPS Agreement, 78, 82, 87, 90n7
TBT Agreement, 78
TRIPS Agreement, 174–75, 179
WTO, 112
biosafety and trade regimes. *See also* biosafety governance; genetically modified organisms (GMOs); "living modified organisms" (LMOs); trade
institutional interplay between, 57–67, 58t3.1

199

200 INDEX

biosafety and trade regimes (cont.)
   interactions, direction and type, 57–59
   non-compliance, punitive sanctions for, 37–38
   performance-interplay and regime principles and rules, 59–62, 60t3.2
   structural linkages and regime objectives, 63–65, 65t3.3
   transboundary movement in GMOs, 65t3.3
Biosafety Clearing-House, 29, 37, 66, 78, 82, 87
biosafety governance
   agricultural commodities, global trade obligations, 33–35
   biosafety, defined, 4
   biotechnology, defined, 15n5
   Cartagena Protocol on Biosafety, 4, 10–12, 26–33
   case of, 8–12
   compliance mechanisms, 36–38
   dispute settlement, 36–38, 41n1, 67
   evolution of, 33–39
   framework, emergence of, 21–22
   global rule diffusion and capacity building, 36
   global rule transmission, domestic, 35–39
   issues, key, 19–21
   membership and participation, 38–39
   relevance of, 39–41
   SPS Agreement and related global regimes, 22–26
Biosafety Protocol. *See* Cartagena Protocol on Biosafety
biotechnology. *See* biosafety governance; genetically modified organisms (GMOs)
Brazil, 34, 38–39, 41n3
Bretton Woods system, 55, 140

Canada
   "beef hormone" conflict, 21
   biotechnology regulation, 106
   Cartagena Protocol on Biosafety, not ratified, 16, 38, 66
   Convention on Biological Diversity, *not party to,* 16
   EU GMO regulations, 22, 37
   exporter of GMOs/LMOs, 35, 41n3, 63–64, 108
   Miami Group member, 27

   retaliatory tariffs on EU, 176
   transgenic crops, commercializing, 27, 38
Carbon Management Research Activity (CMRA), 51
Cartagena Protocol on Biosafety. *See also* Convention on Biological Diversity (CBD)
   advance informed agreement (AIA), 11, 22, 26–29, 66, 78, 108, 144
   Annex III, 11, 82, 109
   Annex III.6, 82
   Article 1, 60t3.2, 65t3.3, 68t3.3
   Article 7, 61
   Article 8, 61
   Article 10, 108
   Article 10.1, 82
   Article 10.6, 31t2.1, 42n4, 59, 82, 90n13, 109
   Article 10.8, 60t3.2
   Article 11.4, 62
   Article 11.6, 108
   Article 11.8, 31t2.1, 42n4, 59, 62, 82, 90n13, 109
   Article 15, 29, 61, 82, 108–9
   Article 15.1, 82
   Article 15.3, 61, 82
   Article 19.3, 27
   Article 20, 82
   Article 26, 84
   Article 26.1, 31t2.1, 109
   Articles 4-7, 28
   biodiversity, 30, 31t2.1, 64, 65t3.3, 82, 87, 90n7, 90n13, 91, 109, 112, 141, 145, 163, xi
   Biosafety Clearing-House, 29, 37, 66, 78, 82, 87
   biosafety governance, 4, 10–12, 26–33
   bulk agricultural commodity shipments, 27, 33–35
   capacity building as vehicle of global rule diffusion, 36
   Compliance Committee and WTO, 117–18
   compliance mechanism, 35, 37–38, 62, 96, 117–18
   "constructive ambiguity", 145
   Convention on Biological Diversity (CBD), 19, 26–27, 78
   countries not ratifying, 16, 39, 66, 116–17, 146
   "creative ambiguity," 12

future action under, 118–19
genetically modified organisms (GMOs), 19
GM labelling imperatives, domestic, 34–35
goals, 50
importer decisions, 31t2.1
importer perspective, 50, 87
international agreements, relationship with, 110–11
jurisdictional delimitation, stepwise, 111–14
"living modified organisms" (LMOs), 10
LMO, "intentional introduction" of, 144
LMO-FFPs, 29, 33, 156n4
LMOs, 28–30, 33, 35, 61, 64, 65t3.3, 66, 108, 112–14, 116–19, 144, 148, 152
"may contain" LMO obligation, 29, 34–35
membership, regime, 63
negotiating rationale and objectives, 26–33
objectives, regime, 64, 65t3.3, 78
pharmaceuticals, LMO-based, 11, 28, 142–43, 155n4
Preamble, 60, 79–80
precautionary language, 30, 32–33, 42n4, 59, 60t3.2, 62, 78, 82–86, 88, 90n13, 109, 146
principles and rules, 59–62, 60t3.2
ratification of, 26
risk assessment and precaution, 108–10
savings clause, 32, 90n11, 110, 145
scientific risk assessment, 29–30, 31t2.1, 42n4, 59–60, 60t3.2, 62, 78, 80, 82–83, 88–89, 90n7, 109, 113, 143–44
socio-economic factors, 30, 31t2.1, 108–10
SPS Agreement, future behavioural interaction, 114–15
SPS Agreement, future interactions and policy implications, 114–19
TBT Agreement, 85–86
transboundary movements of GMOs, 10–12, 59–62, 60t3.2, 67
WTO, future behavioural interaction, 114–15
WTO, future interactions and policy implications, 114–19
WTO, institutional interplay, 140–45
WTO, politics of institutional interplay, 146–48

WTO, resolving clash with, 148–52
WTO influence on, 107–14
CBD. *See* Convention on Biological Diversity (CBD)
Charter of the International Trade Organization, 163
Chicago Climate Exchange and California Climate Action Registry, 58t3.1
Chile, 27, 108
China, 40, 41n3, 42n6, 107, 175–76, 178
CMRA. *See* Carbon Management Research Activity (CMRA)
Codex Alimentarius Commission
  Ad Hoc Intergovernmental Task Force on Foods Derived from Biotechnology, 25
  Committee on General Principles, 66
  domestic regulations, harmonizing, 26
  food safety standards, 9, 20, 25–26, 33, 88
  GM labelling, domestic, 35
  membership, regime, 42n7, 63
  objectives, regime, 64, 65t3.3
  plant health standards, 9
  risk analysis, principles of, 25–26
  secret ballot on safety standards, 25–26
  SPS Agreement, 23, 143, 172
  trade, 155n1
  transboundary GMOs, 66
World Health Organization (WHO), 20
World Trade Organization (WTO), 116
Committee on Trade and Environment (CTE), 144, 149, 163, 168, 177, 179
Consultative Group on International Agricultural Research, 153
Convention for the Abolition of Import and Export Prohibitions and Restrictions, 162
Convention on Biological Diversity (CBD), 4, 10–12, 153. *See also* Cartagena Protocol on Biosafety
  Article 1, 10
  Article 2.2, 78
  Article 19.3, 10
  biodiversity, 26–27, 50, 71, 78, 94, 132, 141, 146, 153–54, 175, 179, 185n24
  Cartagena Protocol on Biosafety, 19, 26–27, 78
  countries not party to, 16, 39, 116–17, 146
  dispute settlement mechanism, 37
  non-ratified by United States, 146
  TRIPS Agreement and, 179

Convention on International Trade in
 Endangered Species (CITES), 7, 53–55,
 86, 100, 163
Convention on Long-Range Transboundary
 Air Pollution (LRTAP), 56
Convention Relating to the Simplification of
 Custom Formalities, 162
COP/MOP I. *See* Meeting of the Parties
 (COP/MOP I)
CTE. *See* Committee on Trade and
 Environment (CTE)

EU Emissions Trading Scheme, 58t3.1
European Union
 GM regulations, 20–23, 25, 28–29, 32–34,
  37–38, 42n6, 108, 147–48, 171
 GMOs, conflict with US over, 20–21,
  36–37, 39, 85, 176

Food and Agriculture Organization (FAO),
 20
Forest Stewardship Council, 54, 58t3.1
Fridtjof Nansen Institute, 6–7, 13

General Agreement on Tariffs and Trade
 (GATT), 71, 105, 132, 163
 Article I, 77
 Article III, 77
 Article XI.2, 174
 Article XIV(e), 174
 Article XVI, 174
 Article XX, 77, 86, 142, 145, 167, 169–70
 Article XX(b), 169
 Article XX(g), 170
 biodiversity, 105
 environmental rights as services, 174
 General Exception, 170
 generalized system of preferences (GSP),
  170–71
genetically modified organisms (GMOs).
 See also "beef hormone" dispute;
 biodiversity; biosafety and trade
 regimes; biosafety governance; "living
 modified organisms" (LMOs); trade;
 transboundary movement in GMOs
 advance informed agreement (AIA), 11,
  22, 26–29, 61, 66, 78, 108, 144
 Africa Group, 34
 anthropogenic drivers on planetary
  support systems, 151
 biological diversity issue, 26–27, 30,
  31t2.1, 64, 82, 105, 109, 132, 139, 141,
  145, 147, 151, 153–54
 Central and East European countries, 28
 Codex Ad Hoc Intergovernmental Task
  Force on Foods Derived from
  Biotechnology, 25
 Compromise Group, 27–28
 defined, 8, 67n4
 documentation obligations, 38
 domestic labelling imperatives, 34–35, 147
 environmental exception for
  conservation, 174
 European Union conflict with US, 20–21,
  36–37, 39, 85, 176
 European Union's GM regulations,
  20–23, 25, 28–29, 32–34, 37–38, 42n6,
  108, 147–48, 171
 Global Industry Coalition, 28, 44
 Like-Minded Group, 27–28, 108, 147
 as "living modified organisms", 27
 Miami Group, 27–29, 32, 34, 38, 85, 108,
  110, 147–48
 moratorium on approval of new GMOs,
  16
 "precautionary principle", 11, 30, 32,
  42n4, 59, 62, 67n3, 83–84, 143, 172,
  185n21
 "prior informed consent", 21, 27, 144
 rules and norms for trade, establishing,
  19–21
 scientific risk assessment, 25, 56–57, 59,
  60t3.2, 90n12, 143–44, 151
 scientific uncertainty, 23, 29–30, 59–62,
  80–81, 83, 89, 90n7, 90n12, 141, 149
 SPS Agreement and related global
  regimes, 22–26
GMOs. *See* genetically modified organisms
 (GMOs)
Group on Environmental Measures and
 International Trade, 163

Havana Charter, 163
Helsinki Convention for the protection of
 the Baltic Sea, 97

IDGEC. *See* Institutional Dimensions of
 Global Environmental Change
 (IDGEC)
IHDP. *See* International Human
 Dimensions Programme on Global
 Environmental Change (IHDP)

INDEX 203

The Institutional Dimensions of Global Environmental Change (IDGEC) Project. *See also* overlapping regimes
  biosafety and trade regimes, 57–67, 58t3.1
  biotechnology, 49–50
  boundary conditions, 5
  causal mechanisms of, 98–99, 99f5.1
  cognitive interaction, 7
  core aspects of regimes, 6, 74, 76, 83
  dependence, forms of, 55–57
  expectations, theory-driven, 134–40
  insights from, deriving, 131–32
  institutions, clustered, 56, 58t3.1, 65t3.3, 73
  institutions, defined, 15n3
  institutions, embedded, 55, 58t3.1, 73
  institutions, nested, 56, 58t3.1, 73
  institutions, overlapping, 56–57, 58t3.1, 65t3.3, 73, 83
  interaction, cognitive, 13
  interaction, directions and types of, 53–54
  interaction, impact-level, 14
  interaction by commitment, 13
  interplay, asymmetrical, 55
  interplay, behavioural, 14
  interplay, bio-physical, 58t3.1
  interplay, deep, 14, 137–39, 139f6.1
  interplay, defined, 3, 52
  interplay, diffusion, 6
  interplay, embedded, 5
  interplay, functional, 5, 53, 57, 58t3.1
  interplay, goal based, 58t3.1
  interplay, hard, 7
  interplay, horizontal, 5, 53, 55, 57, 58t3.1
  interplay, ideational, 6
  interplay, institutional, 50–51
  interplay, intended, 14, 58t3.1, 136, 138, 139f6.1
  interplay, issue based, 57, 58t3.1
  interplay, nested, 5
  interplay, normative, 6, 13, 75–77, 83–84, 88–89
  interplay, operational, 6, 13, 77, 88–89
  interplay, overlapping, 5
  interplay, performance, 59–62, 60t3.2
  interplay, political, 5, 53–54, 57, 58t3.1, 76, 83, 146–48
  interplay, political spillover, 6
  interplay, power based, 58t3.1
  interplay, power-based political, 13, 54

interplay, procedural, 83
interplay, reciprocal, 55, 57, 58t3.1, 63
interplay, shallow, 14, 137–38, 139f6.1
interplay, socio-economic, 57, 58t3.1
interplay, soft, 7
interplay, somewhat reciprocal, 57
interplay, synergetic, 83, 89, 136, 155
interplay, typology of, 52–57
interplay, unidirectional, 55, 57, 58t3.1
interplay, unintended, 58t3.1, 135, 137, 139f6.1, 155
interplay, utilitarian, 6
interplay, vertical, 5, 53, 55
interplay clustered, 5
interplay-performance relationship, 52–53, 59–62, 60t3.2
methodology note, 152–54
modalities, 6
norms, general, 6
problems, 51
research agenda, 52f3.1
research project, international, 4
rules, programmatic, 6
rules, regulatory, 6, 84
rules, specific, 6
Science Plan, 134
secondary aspects of regimes, 6
structural linkage, categories, 55–57
study of, 5–8
taxonomy, limits to, 133–34
time dimension, 7
turf wars, 13, 74, 84, 88
"what is to be done?", 145–46
WTO and Cartagena Protocol, politics of, 146–48
WTO and Cartagena Protocol, resolving clash between, 148–52
WTO and Cartagena Protocol case, 140–45
institutions. *See* The Institutional Dimensions of Global Environmental Change (IDGEC) Project
Inter-Parliamentary Union, 165
Intergovernmental Forum on Forests, 6
International Convention for the Protection of New Varieties of Plants, 153
International Court of Justice, 180
International Human Dimensions Programme on Global Environmental Change (IHDP), 4
International Monetary Fund, 55

International North Sea Conferences, 102
International Office of Epizootics' animal health standards, 9, 23, 172
International Plant Protection Convention's plant health standards, 9, 23, 172
International Treaty on Plant Genetic Resources for Food and Agriculture (ITPGRFA), 6, 153
interplay. *See* The Institutional Dimensions of Global Environmental Change (IDGEC) Project
Interpol, 100
ITPGRFA. *See* International Treaty on Plant Genetic Resources for Food and Agriculture (ITPGRFA)

Japan, 27–28, 42n6, 54, 58t3.1, 106
Japan Agricultural Products Case, 81

Krueger, Anne O., 164, 188
Kyoto Protocol, 53–54, 58t3.1, 97, 100

"living modified organisms" (LMOs). *See also* biodiversity; biosafety and trade regimes; genetically modified organisms (GMOs)
  bulk agricultural commodity shipments, 27, 33–35
  Cartagena Protocol on Biosafety, 28–30, 33, 35, 61, 64, 65t3.3, 66, 108, 112–14, 116–19, 144, 148, 152
  for food, feed or processing (LMO-FFPs), 29, 33, 156n4
  "may contain" LMO declaration, 29, 34–35
  mono-crops, 50
  Monsanto patenting, 50, 148
  SPS Agreement, 57, 61, 108, 111
  transboundary movement of, 10
  transgenic varieties, 29, 33, 41n3, 49
  types of, 20, 29
  World Trade Organization, 106, 108, 111–12, 115–19, 148
LMOs for food, feed or processing (LMO-FFPs), 29, 33, 156n4
London Dumping Convention, 102
LRTAP. *See* Convention on Long-Range Transboundary Air Pollution (LRTAP)

Meeting of the Parties (COP/MOP I), 62
Mexico, 27, 34–35, 38–40, 41n3

Montreal Protocol on Substances that Deplete the Ozone Layer, 38, 53, 56, 58t3.1, 62, 86, 97–98, 100, 103, 152, 178
multilateral environmental agreements (MEAs), 21

NAFTA. *See* North American Free Trade Agreement (NAFTA)
New Zealand, 27, 34, 38–39, 42n6
North American Free Trade Agreement (NAFTA), 35, 53–54
  Article 104, 86
Norway, 4, 8, 27

Organisation for Economic Co-operation and Development (OECD), 9, 27, 97, 105
overlapping regimes
  analysing, 72–73
  aspects, core, 6, 74, 76, 83
  aspects, secondary, 74
  Cartagena Protocol on Biosafety, 78–79
  causal mechanisms, 75–77
  causal pathways, defined, 75
  causal pathways and effectiveness of overlapping regimes, 83–87
  decision-making process of target institutions, 99–102
  disentangling interactions, 94–96
  forum shopping, 13, 74, 85
  institutional interactions, conceptualizing, 96–98
  institutional overlap and regulatory measures, 80–83
  institutional overlaps, 73–74
  interaction, impact-level, 103
  interaction, WTO and biosafety regime, 104–14
  interaction affecting implementation and effectiveness, 102–3
  interaction between nested institutions, 101–2
  interaction through commitment, 100–101
  interactions, cognitive, 100
  overlap and rules of precedence, 79–80
  rules, programmatic, 74
  rules, regulatory, 74
  single-cause effect between two institutions, 96–98
  turf wars, 13, 74, 84, 88

INDEX 205

WTO agreements, 77–78
WTO and biosafety regime, 104–7
WTO and Cartagena Protocol, future interactions, 114–19
WTO influence on Cartagena Protocol, 107–14

PEEZ. *See* Performance of Exclusive Economic Zones (PEEZ)
PEF. *See* Political Economy of Tropical and Boreal Forests (PEF)
Performance of Exclusive Economic Zones (PEEZ), 51
Political Economy of Tropical and Boreal Forests
(PEF), 51
"prior informed consent", 21, 27, 144

Rio Declaration on Environment and Development
Principle 12, 168
Principle 15, 10, 42n4, 59, 60t3.2, 78
Rosendal, Kristin, 7
Rotterdam Convention, 21, 40, 41n2, 45
Russia, 42n6, 54, 58t3.1

SCM Agreement. *See* Agreement on Subsidies and Countervailing Measures (SCM Agreement)
Singapore, 16n6, 27
South Africa, 40, 41n3
South Korea, 27, 42n6
Spaceship Earth (Ward), 183
SPS Agreement. *See also* World Trade Organization (WTO)
Agreement on Agriculture, 173
Agreement on Subsidies and Countervailing Measures (SCM Agreement), 173, 189
Annex A.4, 80–81
Article 2.0, 23
Article 2.2, 80
Article 3.1, 81
Article 3.3, 82
Article 3.4, 82
Article 4.1, 81
Article 5, 78, 81
Article 5.1, 80
Article 5.3, 24, 31t2.1
Article 5.4, 81
Article 5.5, 81, 172

Article 5.7, 23–24, 31t2.1, 59, 60t3.2, 61–62, 87–88, 106, 109, 172
Battle of the Sexes, 113
biodiversity, 78, 82, 87, 90n7
biosafety governance, 22–26
Cartagena Protocol, future behavioural interaction, 114–15
Cartagena Protocol, future interactions and policy implications, 114–19
Codex Alimentarius Commission, 23, 143, 172
domestic GM labelling imperatives, 34
exporter perspective, 50, 87
import restrictions, 105
importer decisions, 31t2.1
International Office of Epizootics, 172
International Plant Protection Convention, 172
jurisdictional delimitation, stepwise, 111–14
LMOs, 57, 61, 108, 111
membership, regime, 63
negotiating rationale and objectives, 22–26
objectives, 22, 63–64, 65t3.3, 77–78
Preamble, 23, 32, 65t3.3
precautionary approach, 62, 81, 83, 86–88
principles and rules, 59–62, 60t3.2
regimes, related, 22–26
scientific risk assessment, 30, 31t2.1, 59–60, 60t3.2, 61–62, 65t3.3, 78, 80–82, 86, 88, 90n7, 108, 143–44, 172, 182
socio-economic factors, 30, 31t2.1
TBT Agreement, 82, 90n9
trade agreement, 77
transboundary movements of GMOs, 59–62, 60t3.2
WTO agreement, 10, 12, 87
Stockholm Convention on Persistent Organic Pollutants, 178
Switzerland, 27–28

TBT. *See* Technical Barriers to Trade (TBT) Agreement
Technical Barriers to Trade (TBT) Agreement
Article 2.2, 78, 90n13, 171
Article 2.4, 171
Article 11.1, 172
Article 11.3.2, 172
Article 12.4, 172

206 INDEX

Technical Barriers to Trade (TBT) Agreement (cont.)
  biodiversity, 78
  Cartagena Protocol on Biosafety, 85–86
  domestic technical standards and regulations, 24–25
  objectives, regime, 64, 65t3.3, 78
  Preamble, 65t3.3
  SPS Agreement, 82, 90n9
  World Trade Organization (WTO), 10, 20
TNCs. *See* transnational corporations (TNCs)
trade. *See also* biosafety and trade regimes; North American Free Trade Agreement (NAFTA); Uruguay Round of Multilateral Trade Negotiations; World Trade Organization (WTO)
  AIA procedure, exemption from, 109
  arrangements, cap and, 134
  balancing biosafety and, 114–15, 120, 142, 145, 147
  barriers to, 140
  Codex Alimentarius Commission, 155n1
  Committee on Trade and Environment (CTE), 144, 149
  compliance measures, 176
  conflicts over GMO, 119, 121n7
  domestic action affecting, 114
  in endangered species, 144
  environment and, 162–66
  free, 103, 113–16, 138, 140, 142–43
  free, in GMOs, 106, 108, 115
  GMOs, documenting, 149
  in GMOs/LMOs, 105, 107, 112, 114, 146–47, 153, 155n4
  in goods/products, 169–73
  indirect restraints on, 141
  in intellectual property rights, 174–75
  interests, environmental and, 104, 115, 150
  interests, privileged, 118
  interests of GMO exporters, 105
  law, 165
  liberalizing international, 102, 141, 183
  negotiations, 116, 142, 176–79
  protection, environment and, 113
  regimes, international, 137–38, 140, 142
  regimes, multiple, 139
  regulation of international, 95
  regulation of trade in GMOs/LMOs, 96, 106, 111–12, 119–20
  restricted, in GMOs/LMOs, 105, 107, 116–18, 146, 150
  restrictions based on scientific assessment, 141, 182
  restrictions by Interpol, 100
  restrictions of multilateral environmental agreements, 116, 144
  risks with GMOs/LMOs, 114
  sanctions, 115, 117, 176
  in services, 173–74
  world, 116–18, 120
  WTO as environmental and trade agency, 165–69, 181–82
  WTO as regime governing international trade, 140, 143
  WTO as trade liberalization agency, 164
  WTO dispute settlement mechanism, 116
Trade and Environment at the WTO (WTO), 164
Trade Prohibitions Convention, 162
Trade-Related Aspects of Intellectual Property Rights (TRIPS), 6, 153, 174–75, 179, 185n24
transboundary movement in GMOs. *See also* genetically modified organisms (GMOs)
  biosafety and trade regimes, 65t3.3
  Cartagena Protocol on Biosafety, 10–12, 59–62, 60t3.2, 67
  Codex Alimentarius Commission, 66
  principles of governing regimes, 59–62, 60t3.2
  SPS Agreement, 59–62, 60t3.2
transnational corporations (TNCs), 150–51
tributes to Konrad von Moltke
  Charbovitz, Steve, 161–62, 197
  Cylke, Owen, 195–96
  Esty, Dan C., 193
  Haigh, Nigel, 194
  Halle, Mark, 193–94
  Kraemer, R. Andreas, 196
  Lavoux, Thierry, 198
  Najam, Adil, 192–93
  Tarasofsky, Richard G., 197
  Weizsaecker, Ernst von, 193
  Young, Oran R., 195
TRIPS. *See* Trade-Related Aspects of Intellectual Property Rights (TRIPS)
turf wars, 13, 74, 84, 88

INDEX 207

UN Fish Stocks Agreement, 7
UNCED. *See* United Nations Conference on Environment and Development (UNCED)
UNEP. *See* United Nations Environment Programme (UNEP)
UNFCCC. *See* United Nations Framework Convention on Climate Change (UNFCCC)
UNIDO. *See* United Nations Industrial Development Organization (UNIDO)
United Nations
   Conference on Environment and Development (UNCED), 166
   Conference on the Human Environment, 163
   Convention on the Law of the Sea, 56, 75, 97
   Environment Programme (UNEP), 9, 105, 175, 182
   Food and Agriculture Organization (FAO), 9, 20, 105, 151
   Framework Convention on Climate Change (UNFCCC), 6, 58t3.1, 97
   Industrial Development Organization (UNIDO), 9, 105
   University (UNU), 8, 161
United States
   "advance informed agreement", 27
   "beef hormone" dispute, 21, 25–26, 36, 41n1, 85
   biotechnology technology, 19, 106
   Cartagena Protocol on Biosafety, *not ratified*, 16, 39, 66, 116–17, 146
   Convention on Biological Diversity, *not party to*, 16, 39, 116–17, 146
   EU GMO regulations, 22, 37, 39, 148, 155n5
   exporter of GMOs/LMOs, 16, 35, 38, 41n3, 63–64, 146
   GMOs, opposes restrictions on, 108
   labelling of GMOs, resists, 147
   Miami Group member, 27
   "precautionary principle" to GMOs, 11, 30, 32, 42n4, 59, 62, 67n3, 83–84, 143, 172, 185n21
   retaliatory tariffs on EU, 176
   secret ballot on Codex safety standards, 25–26
   trade conflict with EU, 20–23
   transgenic crops, commercializing, 27, 38

WTO biotechnology regulation, 106
WTO trade rules on GMOs/LMOs, 147
Uruguay, 27, 41n3
Uruguay Round of Multilateral Trade Negotiations, 108, 141, 144, 149, 164, 168

Vienna Convention, 56
von Moltke, Konrad. *See* tributes to Konrad von Moltke

Ward, Barbara, 183
WHO. *See* World Health Organization (WHO)
World Bank, 165
World Bank Group, 55
World Commission on Dams, 151
World Customs Organization, 100
World Health Organization (WHO), 105, 108
   biosafety governance, 9
   Codex Alimentarius Commission, 20
   International Court of Justice, 180
World Intellectual Property Organization, 181
World Trade Organization (WTO). *See also* SPS Agreement
   as "an environmental agency," 182–83, 184n3
   Anti-Dumping Agreement, 182
   anti-WTO riots, 142
   Appellate Body, 41n1, 61, 81, 85, 167–68, 170–72, 174, 180
   Article 5.1, 182
   Article II, 65t3.3
   Article XII.1, 175
   Article XVI, 182
   biodiversity, 112
   biosafety governance, 4, 10–12, 104–14
   biotechnology regulation, 106
   Cartagena Protocol, future interactions and policy implications, 114–19
   Cartagena Protocol, influence on, 107–14
   Cartagena Protocol, institutional interplay, 140–45
   Cartagena Protocol, politics of institutional interplay, 146–48
   Cartagena Protocol, resolving clash with, 148–52
   Cartagena Protocol Compliance Committee, 117–18

World Trade Organization (WTO) (cont.)
  Codex Alimentarius Commission, 116
  Committee on Trade and Environment (CTE), 144, 149, 163, 168, 177, 179
  dispute settlement mechanism (DSM), 24, 26, 33, 36–37, 41n1, 62, 77, 86–87, 96, 105, 108–10, 115–18, 161, 167, 171, 175–76
  Doha Ministerial Conference, 66, 90n10, 165, 179–80, 185n25
  Doha Round of trade negotiations, 116, 150, 162, 165, 176–79, 183
  environment in WTO negotiations, 176–79
  as environmental and trade agency, 165–69, 181–82
  environmental goods/services in WTO negotiations, 177
  environmental reviews, 179
  "Final Declaration" of the Parliamentary Conference, 165
  fishing in WTO negotiations, 176–77
  future actions within, 115–17
  General Agreement on Trade in Services (GATS), 173–74
  "green provisions", 168
  intellectual property, 174–75
  International Plant Protection Convention, 116
  Johannesburg Plan of Implementation, 179
  Korea Beef case, 170
  LMOs, 106, 108, 111–12, 115–19, 148
  Meeting of the Parties (COP/MOP I), 62
  Ministerial Conference, 106
  multilateral environmental agreements (MEAs), 178–79
  objectives, regime, 63, 65t3.3
  paradigm, towards a new, 179–82
  Preamble to Agreement establishing, 166–67
  as regime governing international trade, 140, 143
  Shrimp case, 167–68
  structural provisions, 175–76
  sustainable development, 167–68, 179–80, 182, 184n11
  Technical Barriers to Trade (TBT) Agreement, 10, 20, 132
  trade agreements, 10, 140–41, 168
  trade-environment linkage, background, 162–64
  trade in goods, 169–73
  trade in services, 173–74
  trade liberalization agency, 164
  as trade liberalization agency, 164
  trade rules on GMOs/LMOs, 147
  TRIPS and biodiversity, 174–75, 179
  win-win-win negotiations, 177–78
WTO. *See* World Trade Organization (WTO)
The WTO as an International Organization (Krueger), 164, 188